The Coldest Crucible

The Coldest Crucible

Arctic Exploration and American Culture

MICHAEL F. ROBINSON

The University of Chicago Press

CHICAGO AND LONDON

MICHAEL F. ROBINSON is assistant professor of
history at Hillyer College, University of Hartford.

The University of Chicago Press, Chicago 60637
The University of Chicago Press, Ltd., London
© 2006 by The University of Chicago
All rights reserved. Published 2006
Printed in the United States of America
15 14 13 12 11 10 09 08 07 06   1 2 3 4 5

ISBN: 0-226-72184-1 (cloth)

Library of Congress Cataloging-in-Publication Data

Robinson, Michael F. (Michael Frederick), 1966–
    The coldest crucible : Arctic exploration and American
culture / Michael F. Robinson.
        p. cm.
    Includes bibliographical references and index.
    ISBN 0-226-72184-1 (cloth : alk. paper)
    1. Arctic regions—Discovery and exploration—American.   2. Explorers—
United States—History—19th century.   3. Scientists—United States—History—
19th century.   4. Science—United States—History—19th century.   I. Title.
    G630.A5R63 2006
    910′.9163′2—dc22

                                                                2005023250

For my parents, Robert C. Robinson and Lucille A. Robinson

Contents

Illustrations

Acknowledgments

I CAN trace, as I expect most writers can, a genealogy of my project, from first ideas to finished manuscript. I have never written it down, but it exists as a set of memories, associated with specific passages from the book. In chapter 5, Walter Wellman sails aloft in his motor-balloon *America*, an event that recalls mornings at the Research Fellows office at the National Museum of American History. The long funeral cortège of Elisha Kane, which threads its way across the United States in chapter 2, conjures up an image of the vaulted ceiling of the old Boston Public Library, which I stared at (from seat 94 below) searching for words adequate to describe the event. In short, it is difficult to disentangle the lives of these explorers from my own, since, from 1997 to 2005, my world has become so closely tied to theirs. This seems a little silly to say, since it is hard to find much that connects the perilous lives of explorers to the bookish life of a writer. But there are things in common. Authors appear, on dust jackets and interviews, as solitary creatures, "going it alone" much as we expect explorers did on their long voyages of discovery. But this is a fiction for authors and explorers alike. My research has shown me how dependent explorers were on large networks of people, in accomplishing their goals and giving meaning to their expeditions. Personal experience has shown me that the same is true of book projects, which would not exist without peers, friends, and patrons. Only through the efforts of many people did this project find its way clear of the docks and into the harbor.

Through the course of this project, Lynn Nyhart has served as my friend, critic, and closest counsel. If this study seems to operate as—to adapt a phrase from Edgar Allan Poe—"a theatre particularly my own," it is one whose dramas are guided by her keen insight and vision of history. James Baughman, Jane Camerini, David Grace, Briann Greenfield, Victor Hilts, Timothy Lemire, Craig McConnell, Ronald Numbers, Louise Robbins, Eric Schatzberg, and Susan Schulten read parts or all of this book. It is a different, stronger work for their efforts. This book also owes its existence to Christie Henry of the University of Chicago Press, who remained confident in the project even when it was mired in the limbo of revision. In addition, the book owes several images (of Robert Peary, Josephine Peary, and Elisha Kane) to the generous efforts of Mark Stapp of the University of Chicago Press, Cally Gurley of the Maine Women Writers Collection at the University of New England, and Kimberly Farrington, David Farrington, and Renata Vickrey of Central Connecticut State University. I obtained the cover art of William Bradford's *Morning on the Arctic Ice Fields* with the help of Katherine W. Baumgartner and the permission of Godel & Co. Gallery of New York.

The National Science Foundation, the Smithsonian Institution, the American Philosophical Society, the University of Wisconsin, and the Osher Map Library at the University of Southern Maine have, together, supported much of my eight years of writing and research. They have also served as my advocates. As a postdoctoral fellow at the Osher Map Library at the University of Southern Maine, I benefited from the enthusiasm and expertise of curator Yolanda Theunissen and the library's founder, Dr. Harold Osher. During my stay in Washington as a Smithsonian fellow, I had the excellent fortune of gaining Pamela Henson as an advisor. Working at the Smithsonian is the closest I will ever come as an adult to the experience of running through Disney World as a child, and Pamela was my guide.

It is difficult to thank my parents, Robert and Lucille Robinson, adequately. They gave me the tools to take on this project, and, in the example of their own lives, offered me a model by which to pursue it. Bart and Penny Troy, my father- and mother-in-law, have supported me with their time, labor, and patience. Penny has often come to my rescue, relieving me of my household duties while shooing me out the door to write. Michele Troy, my wife and dearest friend, has taken on this book as her own, juggling it with her own book projects, family commitments, teaching, and the demands of an academic career. She has given me the time to write, to rant, to revise, even when there was simply no time left. My girls, Tess and Isabella, played no role in starting this project, but they have done a marvelous job finishing it: reminding me every day of the reasons why it must end.

INTRODUCTION

———— ✳ ————

IN the early 1880s, Robert Peary confided to a friend that polar exploration appealed to him as a pursuit "free from discussions, from entangle-
ments, from social complications," a field of work where he wanted to "go it
alone." Arctic expeditions, however, were not the lonely affairs that Peary expected. The voyages were crowded, and Peary's, it turned out, would be more
crowded than usual. On his eighth and final expedition, Peary's ship, the
Roosevelt, steamed north with nineteen officers and crew, forty-nine Eskimos
(including seventeen women and ten children), and 246 dogs.[1] Standing in
the middle of the "stench and disorder of the ship," no one would mis-
take Peary for an Arctic Dr. Livingstone. His success as an explorer hinged
less on his physical stamina than on his ability to organize his expeditionary
team. Disembarking on the barren shores of Ellesmere Island, the sixty-eight
members of the Peary expedition set to work establishing a self-sustaining
colony named Hubbardsville after an important patron. Through the winter
of 1908–1909, Americans and Eskimos prepared for Peary's long trek across
the polar sea. Yet not even on this final leg would Peary "go it alone," as
he had once imagined. When he set off across the rough pack ice, Peary
brought up the rear of a long chain of sledge crews that had broken the trail
and laid down provisions in advance.[2]

If Peary's work in the Arctic depended on the cooperation of dozens of
people, back home his preparations had involved a much larger network.
In 1907 and 1908, his days were consumed by the details of the expedition.
He interviewed volunteers for his crew, negotiated prices for provisions and
equipment, and oversaw the preparations of the *Roosevelt* for its voyage. He

1

spent most of his time, though, raising money. The lion's share of funding came from the Peary Arctic Club, a wealthy group of East Coast patrons. To cover the rest of his expenses, he sold rights to his books and articles. He also lectured relentlessly, relying on invitations from organizations across the social spectrum, from well-to-do New York clubs to local YMCAs and women's auxiliaries. These activities served to fund Peary's voyage to the Arctic, but they had another equally important function. They paved the way for the career he would enjoy as a successful explorer back home. That Peary considered this an important aspect of his work is borne out by his expedition diary. More than half of its fifty-nine pages concern plans for his return: profitable publishing contracts, geographic societies' medals of honor, even a design for a great mausoleum (guarded by a bronze Eskimo and a menagerie of Arctic animals) that he hoped would someday entomb him. Although he had once envisioned the life of the polar explorer as one "free ... from social complications," he had come to appreciate such complications as the warp and weft of Arctic campaigns.[3]

Robert Peary's story raises important questions. How did explorers come to see Arctic exploration as meaningful work, so meaningful that they would risk and sometimes lose their lives pursuing it? What compelled middle-class Americans to follow the adventures of explorers with such avid interest? And why did it become so important for explorers to intimate that they worked alone when the work of exploration required the efforts of so many?

Answering these questions demands that we look at Arctic exploration as an activity that unfolded not only in the Arctic but also at home in America. In this book I take up the story of Arctic exploration during the heyday of its popularity in the United States, from 1850 to 1910. During this period more than two dozen expeditions entered the Arctic on voyages of discovery to rescue missing explorers, find the Northwest Passage, and stand at the North Pole. I pay particular attention to the perils confronting explorers south of the Arctic Circle: the struggles to build support for their expeditions before departure, defend their claims on their return, and cast themselves as men worthy of the nation's full attention. In so doing, I paint a new portrait of these men, one that removes them from the icy backdrop of the Arctic and sets them within the local tempests of American cultural life.[4]

Arctic explorers merit this kind of scrutiny because they were men of consequence in the nineteenth century. They captured the attention of the nation's scientific and political elite and shaped the ideas and institutions of geography at a time when it dominated American science. Their voyages kindled the fires of the Romantic imagination as settlements overran the American frontier. These explorers served as the muses for generations of writers including Washington Irving, Edgar Allan Poe, Walt Whitman, Emily Dickinson, Henry David Thoreau, and Jack London. Painters, too, gained inspiration from fur-clad voyagers. Luminous icebergs and lonely ships fill out the scenes of James Hamilton, William Bradford, and the most heralded landscape painter of the era, Frederic Church. Explorers were also the darlings of the world's most powerful publishers, who fell over each other trying to secure rights to their stories. Arctic voyages thrilled—and sometimes galled—those who read about them in personal narratives, newspapers, school geographies, family atlases, and dime novels. Millions paid admission to see explorers narrate their journeys (with noisy retinues of Eskimos and dogs) at public lectures, world's fairs, museums, and massive traveling polar panoramas. Those who managed to miss this deluge still had to weather a blizzard of Arcticana—expedition sheet music, fabric patterns, silverware, buttons, playing cards, postcards, and cigar bands—that settled into the nation's domestic spaces, cluttering the bureaus, mantelpieces, and piano stands of American homes. So although the Arctic never wore the colors of a U.S. state on nineteenth-century maps, it became a national landscape nevertheless. Its glittering hummocks became the setting of American stories. And its bays and capes, like the men who named them, entered the American vernacular.

Yet it is not merely because explorers enjoyed such popularity then that they should be important to us now. It is also because their popularity is revealing of the times in which they lived. Explorers became associated, gloriously and notoriously, with the traits of the nation. In so doing, they gave voice to hopes and fears that seem, at first glance, far removed from the Arctic regions: about the status of the United States as a civilized nation, about threats to its manly character and racial purity, about the blessings of science, about the dangers of progress. The Arctic, in other words, presented a faraway stage on which explorers played out dramas that were unfolding very close to home. What business did these matters have on the decks of icebound ships? They gave meaning to the voyages. Men believed that Arctic exploration touched on issues so important that they were willing to die for the chance to say something about them. They were the threads that, for sixty years, held the fabric of Arctic exploration together.

To appreciate the role it played in American life, we must consider how Americans learned about the distant regions of their world. In particular, we need to understand the relationship between explorers and scientists, two groups who spoke with authority about matters geographical. In the details of these relationships we begin to see the outlines of a bigger story about the roles of science and exploration in American culture. As much as this book chronicles the milestones and missteps of Arctic explorers, telling this bigger story is the first purpose and final concern.

SCIENCE, CHARACTER, AND MANLINESS

The plot of this bigger story is simple, perhaps deceptively so. It recounts the promising marriage of science and Arctic exploration before the Civil War, explores the decades of their troubled union, and ends by detailing their estrangement and separation at the end of the century. One might imagine that scientists, newly professional, frowned on the motley group of men who led the nation's Arctic expeditions. Scientists, we presume, must have abandoned their relationship with explorers because of the latters' shoddy research or lack of proper credentials. Explorers routinely failed scientists on both counts. But these issues did not bring about their break with the scientific community. Instead it was explorers who first grew restive. At a time when the prestige of scientists had never seemed greater, Arctic explorers got, as it were, cold feet.

Understanding why explorers turned away from scientists requires us to look at the forces that first brought them together. In the 1850s, during the early days of Arctic exploration, science played many roles. Explorers and scientists established their relationship on the basis of an implicit but straightforward quid pro quo. Explorers pledged to collect important specimens and data for elite men of science, who, in return, prodded their colleagues and congressmen to give material support to polar exploration. Scientific elites rarely had the funds to support expeditions directly, but they greased the cogs of exploration in other ways such as scavenging instruments and equipment, testifying before congressional committees in hopes of opening federal coffers, and urging the members of scientific societies to donate funds out of pocket.

Science played another vital, if less visible, role for explorers. It gave them credibility as men of character. Character, specifically manly character, was the lifeblood of Arctic campaigns. It infused the narratives of explorers and animated the prose of their supporters. As Joan Shelly Rubin argues in *The Making of Middlebrow Culture*, the concept of character became increasingly important to nineteenth-century Americans as other measures of social status such as family lineage and wealth began to break down.[5] How did science contribute to character? Influenced by eighteenth-century European ideals of progress, many Americans believed that science offered not only a way of understanding the world but also a means of self-improvement. Writers praised science for its practical uses, but they also viewed it as an uplifting activity that edified the men and women who undertook it. Even such a scientific elite as Joseph Henry, first secretary of the Smithsonian Institution, asked his colleagues to remember "the relation of science to the moral part of our nature." The explorer who lacked formal training in science, then, still had good reason to take it—or at least talk about it—seriously. It signaled to scientific patrons and popular audiences alike that he was, in effect, the right kind of man for the job.[6]

The way explorers talked about science, then, often proved critical to the success or failure of their missions. Science's most important function was as a rhetorical tool, as a means of establishing social authority at home. Explorers' scientific discussions only occasionally concerned particular specimens or experiments. Only fitfully did they discuss, publish, or, for that matter, even conduct Arctic research. For this reason, I spend little time discussing the details of their scientific activities in the Arctic. This might strike some as curious. After all, how can a work claiming to tell us something important about the history of science and exploration do so without looking at the details of exploration science? The reason is that science encompasses a broad range of activities that include, but are not limited to, the practices of observation, collection, and experiment. Explorers were "doing science" not only when they hunted down Arctic specimens but also when they lectured about Arctic geography to small-town lyceums or lobbied scientific elites for professional support. Indeed, these latter activities shaped popular ideas about the natural world of the Arctic much more than did the former. The ways explorers used scientific rhetoric may tell us little about research aboard a ship or may fail to enlighten us about Arctic geography, but they tell us a great deal about how science was used in public debates, the messages it conveyed to different audiences, and the way these messages changed over time.[7]

Using the language of science was only one means by which explorers tried to establish their good character. They also embellished their writings and lectures with widely known moral and religious metaphors, describing their voyages as "quests," "crusades," and "pilgrimages." Such metaphors laid the groundwork for depictions of explorers as "knights," "pilgrims," and "martyrs": men who found historical fame (at least in the eyes of their white, mostly Christian audiences) on the strength of their personal character. Although patrons, readers, and lecture-goers retained their varied interests in Arctic expeditions, most could agree that such voyages had value in illustrating the good character of American men under terrible conditions. Thus, in their focus on manly conduct, narratives helped unify different groups in support of Arctic exploration and encouraged them to find value in expeditions even when the stated goals of such voyages, such as research, commerce, or geographical discovery, fell far short of the mark.

In short, these different forms of rhetoric—scientific, manly, and moral—functioned as explorers' most powerful tools because *stories*, more than specimens or scientific observations, constituted the real currency of Arctic exploration. The writings and lectures of the explorers opened the wallets of patrons, whetted the appetites of publishers, and excited the interest of audiences at home. Not all stories were alike or equally effective. But from the first expeditions of the 1850s onward, American explorers seemed to grasp the importance of narratives to their social success and strove to present themselves as figures with whom audiences could identify. This was no easy task. Even among their mostly white middle-class audiences, explorers encountered competing interests and attitudes. Explorers in the 1850s and 1860s, for example, had to appease powerful scientific elites without alienating their less well-educated audiences and readers.

Understanding the way early explorers put their stories together, combining the ingredients of science and manly character, gives us insight into the processes that eventually diminished the role of science in Arctic campaigns. In the 1850s, traits considered essential to science, such as rationality and discipline, also played important roles in defining ideals of manliness. In this sense, scientific rhetoric served to strengthen explorers' status as model American men. But the role played by science in Arctic campaigns diminished over time. Late nineteenth-century explorers appealed to science less often as they found other goals in the Arctic, other patrons back home, and other forms of rhetoric to energize their campaigns. To some degree, the shift reflected the changing economics of exploration. Explorers found wealthy patrons outside of the scientific community who gave them freer reign. Yet it also reflected an important shift in manly ideals among the middle-class whites who constituted explorers' biggest audience. If early explorers wore the mantle of science to display their high status as men, later explorers found that science sometimes clashed with the other vestments of manly character. As fears of overcivilization prompted explorers to portray themselves as muscular, primitive men, they found it more difficult to simultaneously represent themselves as reasoned and dispassionate. This is not to say that the pursuit of science was inconsistent with manliness in the late nineteenth century. Most scientific communities remained male preserves well into the twentieth century. Moreover, turn-of-the-century scientists sometimes took on the popular image of the muscular man successfully. But for explorers operating outside of scientific professions, it often proved easier to abandon the rhetoric of science than to try to reconcile it with other manly traits that proved so compelling to popular audiences.[8]

Yet explorers' new muscular makeover failed, ultimately, to make them into heroes. It could not shield them from the volleys of their critics. Fame smiled on Peary and a few others, but it often came as a consequence of scandal, not of heroic endeavor. On the eve of World War I, which is when we leave this story, these scandals had left Arctic explorers tarnished figures in the eyes of the scientists, publishers, and popular audiences that had once cheered them on. To offer some measure of their fall: after 1911, it is hard to find writers willing to take explorers seriously enough to *condemn* them. Rather, they lived on as objects of parody, lampooned in jokes, caricatured in cartoons and doggerel. When the young American explorer Vilhjalmur Stefansson returned from the Arctic in 1912, the public and the press remained so skeptical of his accomplishments that even his supporters seemed unwilling to associate him with earlier voyagers. "We shall pay the highest compliment we know to STEFANSSON," wrote the *New York Sun*, "by excluding him from the ranks of Arctic explorers altogether."[9]

THE RISE AND FALL OF THE ARCTIC EXPLORER

What had brought explorers so low? The question seems to have an obvious answer. Explorers did it to themselves. They maligned their rivals. They discredited the Arctic natives who helped them. They lied about their accomplishments. For the past thirty years historians of Arctic exploration have made and substantiated these charges, detailing the rich history of explorers' bad behavior. There can be little doubt that, despite the talk of high character, they were made of the same of the same cloth that we are, that they shared our talents for acting, at times, ineptly and ignobly. And it is also clear that their actions—as any newspaper reader of 1911 could tell you—often had terrible consequences.

But attractive as this explanation is, explorers' behavior can only partly account for their fate as public figures. A number of them had bungled— sometimes epically bungled—things in the Arctic. Yet many of these men had, in the early days at least, managed to avoid the gauntlet of public scandal. Others, such as Peary, found themselves praised for qualities that, at other times, were the subjects of scathing attacks. Those eager to put explorers on trial for their misdeeds fail to see that they were judged by a changing and sometimes contradictory set of standards. My purpose is not to rehabilitate or further castigate these men, who lived in a different age under a different set of codes. Rather, it is to look at the codes themselves, to understand the ways they changed, to track their movements in the public trials by which these men were brought to judgment.

To understand this point, consider for a minute an impossible scenario: How might an Arctic explorer from 1850 be judged by the world of 1900? Elisha Kent Kane (the subject of chapter 2) returned home safely from his expedition in 1855. But for the sake of argument, let us say that he slipped into a glacial crevasse and was frozen solid. Discovered in 1906 by Robert Peary, he was thawed, revived, and returned home, very hungry but little the worse for wear. What would Kane make of his homecoming? The large crowds on the docks would not surprise him, nor would the admirers who told him that he was a testament to manly character. When asked, Kane would be delighted to give a lecture about his experiences because he had been so widely praised in his own day as a gifted orator. But the lecture would disappoint. He would arrive at the hall well dressed (where were the furs, they would wonder?). His address would be eloquent and loquacious, filled with subtle arguments about Arctic geography and peppered with rich, Romantic descriptions of the Arctic and its terrible beauty. Yet it would leave the audience unmoved. Only when he talked about his terrible accident would they sit up in their seats. Opening up the floor at the end of his address, Kane would be surprised by the simple questions he fielded from a "modern" audience. How much weight had he lost? What did raw seal taste like? Had he lived like an Eskimo? He would give polite, if perfunctory, answers to these questions, gently steering the discussion back to more weighty matters. Puzzled, the audience would put on their coats and file out.

Kane would have labored to portray his expedition as a forward-looking enterprise, a symbol of American progress, not realizing that his twentieth-century audience was looking backward, viewing these last voyages to the globe's tiny remaining terrae incognitae through the haze of nostalgia. As for himself, Kane would have tried to show the audience that he was a man of culture and science, not merely to show off but to show them that geographical exploration was the stuff of civilization, that it put the United States at the acme of the civilized world. But his audience would have had enough of civilization, living in the midst of towering office buildings and department stores, breathing in coal fumes and the smell of the stockyards;

they would have hoped that Kane would help them escape from civilization. They would have wanted their explorers to take them lower down—not higher up—the scale of nature, to carry them into a primitive world of long ago, when men, they imagined, battled savages, not anarchists and labor unions. Kane's highbrow descriptions and careful dress would have been lost on them. He would have seemed too effete for their liking, too much a figure of the world they already knew. He would not have seemed *touched* enough by his encounter with the wild. The crucible of the Arctic was supposed to test a man's mettle, they would think, not gild it. It was there to burn away the artifices of civilization and reveal the elemental manliness underneath, pure and unalloyed.

The point I am trying to make with this tale is this: Kane would fail to live up to the expectations of a later age, not because of his failures of conduct but because of the changing expectations of his audience. Truth be told, one did not have to fall into a crevasse for fifty years to fall out of step. Explorers faced these shifting expectations about manly character, about the role of science, and about the meaning of civilization all the time. There were other things, too, that explorers could not control: the press's new willingness (indeed, eagerness) to print embarrassing stories, the changing status of scientific and geographical societies, and the role of other regions—and other explorers— in shaping Americans' views of the Arctic. The personal conduct of explorers would provide the spark of controversy, to be sure, but the fate of this spark, whether it died away or flamed into scandal, would ultimately depend on matters that lay beyond their reach.

THE LIMITS OF EMPIRE

Even by nineteenth-century standards, Arctic explorers were conspicuously patriotic. At times, the rhetoric of nationalism became thick enough to ob- scure the Arctic entirely from view. Omitting brief references to the North Pole, one might think that Peary's 1907 address to the National Geographic Society was a political stump speech: "When the wires tell the world that the Stars and Stripes crown the North Pole, every one of us millions from child to centenarian, from laborer and delver in mines, to the 'first gentleman' in the land, will pause for a moment from consideration of his own individual horizon and life interests, to feel prouder and better that he is an American, and by proxy owns the top of the earth."[10]

Explorers advertised their love of country in many ways. They rarely missed an opportunity to name their vessels in ways that could connect their voyages with the nation. Peary's *Roosevelt* was only one example. Other explorers christened their crafts with names such as *America* and *United States* (which, in case anybody missed the point, set sail on the Fourth of July). Explorers unfurled American flags when they reached new lands, coasts, and points "farthest north." They unfurled them in greater numbers back home. The flag became a staple image in Arctic slide shows, book plates, and magazine illustrations from 1850 until 1910. They even waved from the pages of the juvenile literature of exploration. Children reading *The Snow Baby* by Josephine Peary learned of her voyage to northern Greenland with her husband Robert and the birth of daughter Marie at Peary's base camp. They saw little of Robert's journey across the Greenland icecap, but this did not prevent Josephine from hoisting the Stars and Stripes: it framed portraits, draped high chairs, even swaddled little Marie in the arms of her mother (see fig. 1).[11]

But we must be careful how we interpret these professions of patriotic ardor. True, they reflected a spirit of cheerful chauvinism among explorers and their supporters. True, too, they showed the hope of explorers that such patriotic displays would excite readers. Perhaps they were a sign of something deeper, a signal to the Old World that the United States had at last entered its imperial pubescence. But if we stop here, we miss the other subtle message conveyed by this patriotic hyperbole: that explorers were nervous about their niche in American life and their connections to the country. After all, they had few of the perks or protections afforded other explorers. Although they had friends in the U.S. Coast Survey and the Naval Observatory, Arctic explorers had no permanent home in the nation's federal bureaucracy. Most polar voyages were privately organized and funded, unlike the expeditions sent west under the direction of the Army Corps of Engineers or the Pacific expeditions organized by the navy. After the Civil War, large commercial interests such as the *New York Herald* and the *Chicago Record-Herald* took on roles as Arctic patrons, easing the burden on explorers to drum up their own support. But this easy patronage came at a price. No one, not even diehard supporters, believed that newspapers funded these expeditions because of simple patriotism. They were vested interests and, as such, threatened to frame Arctic exploration as a matter of corporate, not national, concern. And so, with funds in hand, explorers still had to show that their love affair with the Arctic really grew out of a deeper love of country. Frenzied flag-waving, then, did not afflict them as might some neurological tic of empire; it was a cultivated response, an effort—earnestly felt, perhaps—to shore up support among those who

FIGURE 1. Josephine Peary swaddles her daughter Marie in northern Greenland. From J. Peary, *Snow Baby*, 18.

did not see Arctic exploration as an obvious or important part of American nation-building. *See NASA*

Some audiences responded to these patriotic appeals. But we should not assume that they therefore viewed the Arctic as vital to national interests. Frankly, the United States had more important imperial projects to worry about. As Americans fielded their first expedition to the Arctic in 1850, the country had become obsessed by a far more pressing territorial question: what to do with the spoils of the Mexican-American War. Specifically, how would the new territories enter the Union, as free states or slave? It was a question that threatened to shatter the Union. The Arctic, in this context, provided a happy distraction. It was a project that could still rally Americans together at a time when they were tearing themselves apart. The pattern repeated itself later in the century: the Arctic was a place that, unlike the Philippines or Latin America, could be explored without being administered, a place to flex imperial muscle without having to do the heavy lifting required by a colonial empire. Because it fed on the martial symbolism of conquest, however, Arctic exploration could never compete with real acts of war. Ersatz battles with nature made for good reading during idle days of peace, but they seemed rather ridiculous when Americans were dying elsewhere in real battles. Even the patriotic Arctic buff knew the difference between the polar regions and the Panama Canal and that explorers were doing imperial theater, not protecting the nation from its enemies. In short, the Arctic lived a vibrant life in the American imagination, but it did so for reasons that were more cultural than imperial. It was a stage for explorers to show the traits of character most cherished by Americans, traits that seemed threatened back home. If the Arctic seemed at times a good place for Americans to show their colors to the world, it was an even better place to calm their fears about themselves.

I HAVE been guided by new scholarship rich in the social and institutional contexts of exploration. The works of Janet Browne, Felix Driver, and Helen Rozwadowski are inhabited not only by explorers but by natives, scientists, patrons, and colonial administrators. These scholars, too, have sought to find new ways of talking about exploration that integrate empire but avoid, in Driver's words, "the imperial will-to-power." It is my tribute to them and others that I use their approaches and try to extend them to the worlds of art, popular literature, and public spectacle. I do so not merely to be comprehensive but because these contexts change the story I tell. The relationship between explorers and scientists only makes sense if we look at Arctic exploration across the broadest vista of American cultural life. Confining our attention to expeditionary journals, scientific correspondence, and congressional debates will not reveal the reasons why Arctic explorers lost interest in scientists at the end of the nineteenth century. The answers arrive only when we consider these artifacts in relation to the polar landscapes, sideshows, cartoons, and editorials that conveyed the messages of Arctic exploration to popular audiences. They carried with them their own ideas and aesthetic vogues, tell us of the other roles that enchanted explorers, and caution us to remember that—even in the age of modern science—embracing modernity and science was a choice, not a self-evident prescription.[12]

The chapters of this book unfold chronologically, focusing on the explorers who most captured the attention of American audiences. Chapter 1 considers the question that anchors the rest of the book: How did the Arctic, a region so geographically removed from Americans, become so culturally important? It considers the Jacksonian Era, when exploration found its niche in American culture and when the Arctic first piqued the interest of politicians, the press, and the public at large. Chapter 2 examines the campaigns of Elisha Kent Kane in the 1850s. It argues that Kane's varied use of scientific language established him as a man of character with popular and elite audiences alike. High praise from these different audiences made it easier for Kane to overcome his poor performance in the Arctic and gain wide acclaim as a scientific hero. Chapter 3 looks at the efforts of Isaac Hayes and Charles Hall to reproduce Kane's popular success and considers why they fell short. Hayes and Hall used different strategies to promote their expeditions, strategies that illuminate growing tensions between manly and scientific ideals of exploration. The 1881 expedition of Adolphus Greely, the subject of chapter 4, was by all accounts the most scientific American polar expedition of the nineteenth century. Yet it only hastened the growing divide between Arctic explorers and the scientific community. Understanding why requires us to look at important changes back home: institutional, in the rise of newspapers as powerful new voices in the Arctic narrative, and social, in middle-class anxiety about the decline of American character. Chapter 5 discusses public optimism regarding mechanical solutions to the "Arctic Problem" at the end of the century by focusing on the "motor-balloon" expeditions of Walter Wellman and the voyages of Peary. Despite widespread enthusiasm for using machines in the Arctic, some explorers and members of the public felt that they diluted the value of such exploration as a test of man against nature. The story of Wellman and Peary, then, reflects broader ambivalence about science, technology, and manly ideals at the turn of the century. Both Peary and Frederick Cook, the subjects of chapter 6, built their personas on a model of manliness that had little to do with science. When controversy engulfed both men's claims to the discovery of the North Pole, they had none of the protections once afforded scientific explorers. Contradictory attitudes about manliness only fanned the flames of the controversy, undermining the role of the Arctic explorer as a cultural hero. The conclusion considers the significance of Arctic exploration to the story of American science. It also considers the spread of "Arctic fever" to other expeditionary projects, a process that has shaped both popular attitudes and federal policies toward exploration in the past century. It constitutes Arctic exploration's longest-lasting cultural legacy.

———— ✳ ————

CHAPTER ONE

Building an Arctic Tradition

EXPLORATION IN THE EARLY REPUBLIC

Citizens of the new United States did not pay much attention to geographical discovery. War with Great Britain had given them other things to think about. Yet war proved to be an important catalyst of exploration because it forced Americans to take stock of the vast territories surrounding them, to consider geographical knowledge as a bulwark that might protect their new republic. Appreciation of geographical knowledge did not begin with the Revolution—colonists had been surveying the American wilderness for decades—but the upheaval greatly enhanced geography's importance. This became clear when the colonies, having successfully cleaved themselves from Great Britain, set about reassembling themselves as a single nation. Both acts—cleaving and reassembling—generated difficult boundary disputes. Great Britain and Spain contested the northern and southern borders of the United States, while states quarreled among themselves about control of the territories ceded by Great Britain east of the Mississippi. The new government resolved the disputes in piecemeal fashion by means of international treaties and the Great Land Ordinances of the 1780s. But it gained something important in the process: practical experience dealing with the murky status of new territory.

In the peace that followed the war, the government found new uses for geographical knowledge. The end of hostilities with Great Britain left the Continental Congress in charge of new lands and huge debts. It followed that dispensing with the former might pay down the latter. Selling the nation's new acreage first demanded surveys. The government appointed Thomas Hutchins in 1785 as Geographer of the United States to initiate a public survey of the western lands. On the banks of the Ohio River, Hutchins established the "Point of Beginning," a mark that anchored a new grid of land title to the American continent. Although Hutchins looked at the new territories with the discerning eye of a naturalist, Congress had no interest in funding a scientific survey. The nation's geographer was to measure nature, not marvel at it. Any gazing at the landscape should have some practical objective, such as the discovery of "mines, salt springs, salt licks, and mill seats." Geographical exploration, then, found early government support, albeit in a form that restricted geography to a very narrow range of inquiry. Expeditions to the frontier (which at this point still lay east of the Mississippi) served to clarify borders, establish lines of property, and identify objects of practical value.[1]

These objectives did not apply to regions farther west. Americans knew almost nothing about the inhabitants, resources, and river systems of lands west of the Mississippi River. The Mississippi that snaked its way down the political map of North America served as a border, neatly dividing the United States and New Spain. Yet the real river joined these regions together, connecting tribes and resources along its vast waterway and tributaries. The edge of the American frontier, in other words, was not a lonely backwater but a great highway that linked the United States to peoples and places unknown. Such a highway posed obvious opportunities and perils. It promised commerce with additional Indian tribes and a possible route to the Pacific Ocean, but it also served as a route of invasion or blockade. Indeed, Spain's cession of the Louisiana Territory to the expansionist regime of Napoleon Bonaparte deeply worried President Thomas Jefferson, who viewed it as an event "very ominous to us." Assessing the many threats and opportunities presented by the Far West thus required expeditions of greater flexibility than the public surveys of the 1780s. Explorers of the western frontier had to do more than look for salt licks and mill seats. They would need to consider geography more broadly, bringing to its assessment a new series of questions.[2]

In 1803 Jefferson wrote a secret letter to Congress explaining the necessity of such an expedition. The goals of the enterprise—contacting Indian tribes, establishing trade links, and finding a water passage to the Pacific Ocean—could all be justified as commercial objectives that would easily fall within the constitutional powers afforded to Congress. Geography and natural history would also have roles to play, not as principal objectives but as possible windfalls of discovery. That the expedition "should incidentally advance the geographical knowledge of our own continent," Jefferson explained, "cannot be but an additional gratification." No one mistook this brief reference to "geographical knowledge," coming at the end of the letter, as a declaration of a new kind of scientific exploration. But it was enough for Jefferson to tuck research into the corners of the mission, outfitting his "Corps of Discovery" with proper scientific equipment and submitting its leader, Meriwether Lewis, to a crash course in scientific study.[3]

In this sense, the Corps of Discovery expedition did represent something new, a U.S. expedition that bundled the study of nature together with commercial and military objectives. In this form it more closely resembled the European discovery expeditions of James Cook and Jean-François de la Pérouse than it did the earlier U.S. surveys. On the surface, it made sense to include scientific objectives in discovery expeditions if only because geographical data could be used for many ends. Measurements of magnetic variation, for example, interested not only navigators and map makers but also scholars who studied terrestrial magnetism. Yet the inclusion of scientific objectives did not simply make good sense in the field. It smoothed out bumps in expedition planning back home. In Europe, for example, discovery expeditions gained the approval of scholarly societies who lent their social prestige to the new voyages. And when the journeys went wrong, as did La Pérouse's expedition to the Pacific, narratives and natural history collections helped ease the national pain of failure.[4]

Not all of these benefits yet applied to exploration in the United States. In 1800 the nation did not possess Europe's powerful scientific societies. Nor did its expeditions yet need scholarly cover for imperial designs. But the Corps of Discovery's directive to advance geographical knowledge, however modest, gave the expedition its first success long before it crossed the Mississippi River. That is, it allowed the United States to present its mission to European powers—who would otherwise disapprove of a U.S. expedition on their soil—as a scholarly adventure that fell outside the realm of politics. "The nation claiming the territory," wrote Jefferson to Congress, "regarding this as a literary pursuit... would not be disposed to view it with jealousy." He showed extraordinary deftness, then, in his campaign to launch the Corps of Discovery expedition, not simply because he understood the value of science as a tool of exploration but because he recognized that science had multivalent properties, that it served different functions in the naturalist's garden, in the halls of Congress, and on the banks of the Missouri.[5]

Ironically, the success of the Lewis and Clark expedition in meeting its objectives has misled us as to its historical significance. The party established good relations with Indian tribes, gathered an extensive natural history collection, and found a route to the Pacific Ocean. These accomplishments have given great luster to the expedition today because they are so unrepresentative of later western journeys. The Corps of Discovery shines brightly to us because it beckons from the far edge of a century so clouded by tragedy: the conquest of the Far West, the exploitation of its natural resources, and the subjugation of its peoples. Yet none of this was obvious to Americans living in the early nineteenth century. They did not know how history would turn out. Nor, for that matter, would they have understood our concern for the welfare of Indians or the American wilderness, forces that from their perspective seemed both vast and threatening. More to the point, they *could not* make these judgments about the expedition because they knew so little about it. The expedition was not widely reported in the popular press, nor was it discussed much among the small scientific societies that concerned themselves with such issues. Few Americans could marvel at the natural history of the Far West because most of the expedition's botanical collection was destroyed in transit to the East Coast. They could not read the journals of Lewis and Clark because they did not appear in print until 1814, and then in curtailed form. Despite its success in the field, then, the Lewis and Clark expedition left few tracks on the wider culture of Jeffersonian America.[6]

Yet it did leave its mark among a smaller, eclectic group of American military officers, naturalists, and artists who found that western exploration held opportunities for professional advancement. When the United States purchased the Louisiana Territory from France in 1803, it doubled the size of the country and gave political urgency to the task of exploration. Although none of the western expeditions to follow Lewis and Clark—those by Zebulon Pike (1805–1806), Thomas Freeman (1805), and Stephen Long (1820)—proved as successful in the field as the Corps of Discovery, they maintained the diversity of exploration as a practice, bringing together a wide array of individuals who used exploration for different ends. For the military officers that led them, expeditions offered rare peacetime commands that positioned them well for future advancement. Young naturalists joined these expeditions not only to gather specimens but also to gain experience and social prestige. Artists such as Titian Peale and George Catlin tagged along, too, depicting the bounties of nature for display and profit back home. Despite the opportunities that exploration presented, however, no single agency of government or the military gained control over American expeditions. During the first half of the century, exploration remained a largely ad hoc activity.[7]

By the late 1840s, however, important aspects of exploration had become institutionalized. Powerful agencies had sprung up in Washington, DC, to assist or manage expeditions, as well as to tend to the collections that they brought back. The Mexican-American War (1846–1848) brought new territories under U.S. dominion and fueled demand for more surveys. For years maritime states had complained of the lack of accurate coast and harbor charts, and their demands had given rise to the Coast Survey in 1816. Although it had foundered for decades, the charisma of its new superintendent, Alexander Dallas Bache, and the need for charts of the country's vastly expanded coastline now brought new luster to the Survey. Exploration also occupied the new Naval Observatory, which was founded in Washington in 1842. Its director, Matthew Fontaine Maury, demonstrated the value of exploration to American commerce by using data from expeditions to produce valuable wind and current charts. Across town the Smithsonian Institution was becoming the national warehouse of exploration by the late 1850s, piling up natural history collections brought back from expeditions to the West and to South America. The Smithsonian's first secretary, Joseph Henry, went to work cataloging these at the same time that he organized field projects of his own. While national expeditions still lived or perished at the whim of Congress, these agencies found important niches for themselves: surveying, collecting, and cataloging the fruits of the American empire.[8]

Increasingly, U.S. expeditions set sail on missions symbolic as well as practical. By the 1830s, Congress was looking to exploration as a means of enhancing the nation's reputation, not simply a way to advance military or commercial objectives. Starting with the departure of the U.S. Exploring Expedition to the Pacific in 1838, the nation pursued a series of international expeditions of discovery. Whereas the government had frowned on all but the most practical goals in exploring the West, it proved more indulgent of scholarly objectives in exploring the world. It was not that the government placed greater significance on the geography outside its borders. Rather, it was that the wider world offered a more prestigious stage for explorers than did the American West, a place where their actions would be more keenly noticed. Under such scrutiny, the expeditions put on their best face, sailing with corps of "scientifics" to advance geographical knowledge, and in the process, to persuade other nations that the United States was more than a republic of untutored farmers. In short, pursuit of knowledge gave U.S. expeditions symbolic heft. It ushered the nation into an Enlightenment tradition of imperial voyaging and, its organizers hoped, into the ranks of civilized nations.[9]

That exploration had such symbolic power came as no surprise to the reading public. Since the late 1700s, Americans had supported a small market for exploration narratives among American publishers. The growth of an educated middle class in the early nineteenth century brought about a sharp rise in book and newspaper publishing houses and fueled greater demand for exploration and travel literature. "Travels sell about the best of anything we can get hold of," remarked the head of Harper and Brothers. It helped that European explorers were already publishing narratives with an eye toward middle-class readers on both sides of the Atlantic. American magazines joined the fray, writing synopses and reviews of the latest narratives received from Europe. Novelists such as Washington Irving, Edgar Allan Poe, and James Fenimore Cooper made expeditions the subject of true-life and fictional works. By the 1830s travel writing had become a genre broad enough to appeal to many interests. It encompassed the erudite writings of explorer-naturalist Alexander von Humboldt and the lurid reports of the *Mariner's Chronicle,* a British collection of "shipwrecks, fires, famines, and other calamities" that spawned fifteen different American editions between 1806 and 1857. Was there anything exceptional about American interest in exploration? At first glance Americans' fascination seems derivative, a taste acquired from Europeans who had already found an appetite for the literature of discovery. On the other hand, many Americans, in particular, Anglo-Americans, felt a certain kinship with explorers who had left the heart of the Old World for places new and dangerous. Stories of exploration resonated with readers around the world, but they struck a special chord with U.S. readers inclined to see them as parables of national history and identity.[10]

The evidence for this comes from many quarters. American writers and artists increasingly used exploration as a vehicle for examining character, both personal and national. Whereas early Puritan writers linked American identity to new institutions of church and state, many nineteenth-century writers such as Irving and Cooper looked for the heart of America outdoors, in the individual's encounter with nature. A similar shift was under way among artists. Early painters sought to illustrate national identity by depicting the events and participants of the American Revolution. By the 1830s, however, a new school of artists had started to root American identity in the land itself. In the 1820s the painter Thomas Cole found his inspirations locally in the sometimes bucolic, sometimes fierce landscapes of the Catskills.

His pupil, Frederic Edwin Church, set off for more distant and difficult locales in the Rockies and the Andes, producing scenes that thrilled East Coast audiences and established him as the preeminent landscape painter in America.[11]

More than fortune drove Church and fellow artists into the wild. They wanted to produce landscapes that moved beyond the conventions of European painting, embodying a distinctly American vision of nature. Old World painters had developed their craft to portray towns, pastures, and ancient ruins. Europe's "once tangled wood," Thomas Cole observed in 1836, "is now a grassy lawn." For American artists schooled in Europe, this lawn seemed too well groomed to teach them much about painting the wilds of America. They looked to the journey west, then, with new purpose. First, it would bring them to the breathtaking places they wanted to paint. But the journey would also prepare them for the task of painting, providing an apprenticeship with nature whereby they learned how to see and appreciate its strangeness. Painters traveled in hopes that the wilderness would get under their skin, alter their perceptions, and infuse their works with something unique. To educated Americans in the middle decades of the nineteenth century, this was not a silly or eccentric project. Church and the roaming artists of the Hudson River School recapitulated a national story. As they sought the frontier, they were being shaped by it. In the early republic, Jefferson had looked to western exploration for knowledge and commerce. But now, in the Age of Jackson, journeys westward signified more than the sum of objects practical and political. They had also become rites of passage from which flowed the wellspring of the American spirit.[12]

THE ARCTIC

The wellspring of that spirit did not yet flow to the Arctic, a region that held little intrinsic meaning for Americans. It was the British who planted the Arctic in the nation's cultural consciousness. Idled by the peace that followed the Napoleonic Wars, the British Admiralty renewed its long search for a northwest passage over the top of North America (see fig. 2). It seemed, at first glance, to be a reasonable project to pursue because a shorter route from the Atlantic to the Pacific would reduce the time and costs associated with voyages around Cape Horn. This goal had launched dozens of voyages since the sixteenth century. Yet few still hoped to reap the benefits of such a northwest passage because earlier voyages had all but confirmed that the passage, if it existed and could be found, would be far too perilous to be used as a commercial sea route. This did not stop John Barrow, the second secretary of the Admiralty, from seizing on Arctic exploration for other ends. In public he spoke of its benefits "for the advancement of geography, navigation, and commerce." Behind the scenes, Barrow conceived of the Arctic as a new theater of war, one in which his ships battled icebergs and pack ice rather than French ships of the line. In the Arctic, he observed, British officers could risk their lives for higher, more civilized ends than those found on the fields of Europe. When whalers returned from the Arctic in 1816 and 1817 reporting that normally ice-choked bays were open, he used the news to launch a series of expeditions the following year. Thus began a thirty-year period of naval and overland exploration that would send thousands of British men into the Arctic.[13]

Almost immediately, the spectrum of British popular literature, from high-brow journals to the penny press, represented the Arctic as a new form of battle and a new source of honor for the Crown. Just as they had shown their character on the fields of Waterloo, military men now revealed it anew in their assault on "the Frost King." Novelists picked up the theme as well. The first and final scenes of *Frankenstein*, which Mary Shelley published just as Barrow's ships set sail, take place in the Arctic, aboard the ship of a British explorer. "Return as heroes who have fought and conquered," Doctor Frankenstein tells wary crewmen, "and who know not what it is to turn their backs on a foe." The narrator of Benjamin Bragg's novel *Voyage to the North Pole* sounds a similarly martial tone: "What man would go to battle if he was afraid to run the chance of wounds and death?" Explorers returned to Britain as if from the battles of war, receiving the honors of military heroes. Expedition leaders such as George Back, John Ross, and William Parry found promotions, knighthood, and membership in presti-gious scientific clubs waiting for them.[14]

The American popular press followed these expeditions closely. Dailies and weeklies remained dependent on Britain as a source of news and culture in the early nineteenth century, and they filled their pages with articles culled from British magazines and newspapers. When the British popular press heaped attention on expeditions to the polar regions, the American press quickly followed suit. Up and down the East Coast, magazines such as the *Athenaeum* (Boston), the *Saturday Magazine* (Philadelphia), and the *Niles National Register* (Baltimore) offered their readers the latest reports about British explorers in the Arctic. Book publishers also took advantage of the new interest in that region. Because the U.S. government did not recognize foreign copyrights, publishers often reprinted (that is, pirated) narratives at

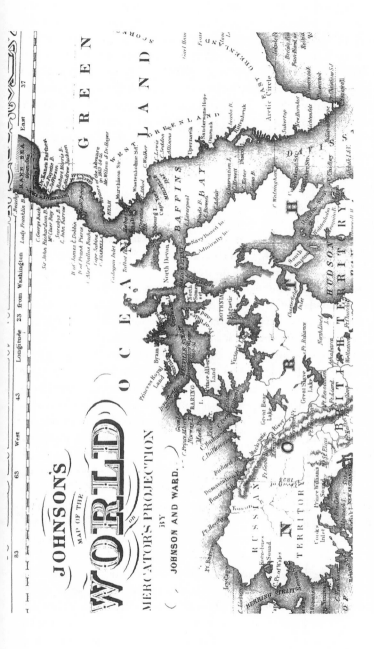

FIGURE 2. This detail from A. J. Johnson, *Johnson's New Illustrated Family Atlas of the World* (New York: A. J. Johnson, 1866) shows the latest Arctic discoveries. Most expeditions launched to find Sir John Franklin in the 1850s headed west through Lancaster Sound (at center, unlabeled). As explorers became more focused on reaching the North Pole, they headed north through ice-choked Smith Sound (top right).

little cost. Harper and Brothers of New York published a number of British accounts as part of their inexpensive Family Library and Boys and Girls Library series. The latter abridged British narratives and gave them moral messages such as those in *Uncle Philip's Conversation with the Children about the Whale Fishery and Polar Seas*, which mined the search for the Northwest Passage for lessons about honorable behavior. In Philadelphia, the powerful firm of Carey and Hart published narratives, atlases, and geography primers, all of which reprinted or incorporated news about the latest expeditions to the Arctic. These were types of exploration literature with which American readers were familiar. Despite the novelty of the Arctic as a site of exploration, then, audiences were quick to accept it as a stage for dramas both sensational and symbolic.[15]

Influenced by this barrage of British literature, Americans began cultivating their own interest in the polar regions. *Godey's Lady's Book* not only reviewed the latest British Arctic narratives but also featured poems about the polar regions such as "The King of Icebergs" and "The Arctic Lover to His Mistress." Explaining the geography of the polar regions became a personal crusade for John Cleves Symmes, a college lecturer. In the same year that Barrow's expeditions left for the Arctic, Symmes published *Symmes's Theory of Concentric Spheres*, which claimed that large openings at the North and South Poles led to a series of cavities inside the earth. Symmes presented his ideas to his students at Union College and at public lectures throughout the 1820s. Few people took his ideas seriously, but they generated considerable interest in the press. (Discussion of his works and allusions to his "Holes at the Poles" continued into the twentieth century.) Symmes's ideas also inspired the American journalist Jeremiah Reynolds, who took up his theory in lectures and in print. Reynolds argued Symmes's views before Congress in 1836, winning a $300,000 appropriation for an expedition to the South Pole, which soon became the U.S. Exploring Expedition. Through it all, America's new interest in the Arctic carried with it an old interest in Europe's good opinion. A principal goal of his voyage to the South Pole, Commander Charles Wilkes told members of Congress, was to make an American contribution to geographical science that would stand up to the accomplishments of Europe.[16]

As talk of an American exploration of the South Pole intensified, the subject soon entered American fiction. *Symzonia* (1820), published under the pen name Captain Adam Seaborne, told the story of such an expedition and passage into the earth's interior. Edgar Allan Poe related a similar, if more macabre, voyage to the South Pole in *The Narrative of Arthur Gordon Pym*

(1837). Other authors such as Cooper and Robert Bird also set a number of stories in the Arctic. Adventure remained a staple of these tales, but patriotic themes, too, found exposition. In the same high-blown language used by Barrow and other British advocates of Arctic exploration, Poe argued in the *Southern Literary Messenger* that the polar expedition "is called for by national dignity and honor." The British showed how polar landscapes— austere, hazardous, and commercially useless—could have national value. The flurry of polar literature in the United States is evidence that many Americans were watching carefully. Eager to rival the British, they had set their expeditionary sights on the opposite end of the earth. In Antarctica American explorers might match the British quest for the Northwest Passage. "Here," Poe urged, "is a wide field open and nearly untouched—'a theatre particularly our own.'"[17]

But events in Britain would again turn American attention to the North. In 1845 the Admiralty sent John Franklin (see fig. 3) into the Arctic with two ships and 129 officers and men to complete the Northwest Passage. Franklin was a veteran of three Arctic expeditions. An overland expedition had brought him to the edge of starvation at the polar sea and fame back in England as "The Man Who Ate His Own Boots." His ships, *Erebus* and *Terror*, with reinforced hulls and steam-powered propellers, had also proved themselves in the polar regions. Thus it was surprising when Franklin did not return the following year or the year after that. In 1848, with no word yet, the Admiralty sent a series of expeditions to look for him, focusing on the northern coast of North America and islands off its shores. They found no signs of the expedition. Lack of news deepened the mystery surrounding the lost party and fueled public interest. By 1849 the British press had become fixated on the Franklin search, and it had prompted wide coverage in American newspapers and magazines as well. Readers wrote letters to the editors offering theories about Franklin's whereabouts, some of which were odd enough to warrant their own stories. A spiritualist alerted the *New York Tribune* that she had located Franklin during a seance. He was alive and well, she reported, held captive in Japan. The discovery amused other New York and New England newspapers enough to reprint the story along with mocking commentaries.[18]

Interest in the Franklin story, already high, soared higher when Jane Franklin, the wife of the missing explorer, appealed directly to President Zachary Taylor for help. Caught in the full glare of the search, Taylor urged Congress to outfit an expedition, a decision that shifted the spotlight from the White House to Capitol Hill. Meanwhile, Jane Franklin had also taken up correspondence with a sympathetic New York shipping merchant, Henry Grinnell (see fig. 4), who pledged his ships for a U.S. search mission. Grinnell promptly circulated a petition to Congress to make his offer official. So when the House and the Senate took the matter up for debate in the spring of 1850, they had to answer two difficult questions. Should the United States field an expedition to find the military party of foreign power? And if so, should they accept the ships of a private citizen in order to do it?[19]

Congress argued about the moral and symbolic merits of an Arctic search expedition. There were, after all, few other justifications for the voyage. Many in Congress found the humanitarian appeal of a rescue mission compelling. Rescuing Franklin exemplified "beneficent design," one that would gain the United States the world's praise as a result. "When this love of glory seeks its accomplishment in noble discovery," reflected Edward Baker of Illinois, "I not only admire, but honor it; and I am honored in being allowed to aid it." But lofty moral arguments became tangled in issues of self-interest. The disappearance of Franklin had cast a spotlight on the Arctic as never before. Few congressmen failed to recognize that an expedition of mercy in the Arctic would gain far more attention than expeditions to less exciting regions of the globe. "I know that national reputation is national property of the highest value," stated Arthur Butler of South Carolina, "and, if we can acquire reputation by making this discovery, I shall rejoice at it."[20]

Adversaries of the search argued that "national reputation" seemed a rather breezy objective for an expedition that entailed real risks. Opposition was strongest among southern Democrats, who had little interest in supporting the initiative of a Whig president. Launching an American expedition to the Arctic, argued Thomas Bayly of Virginia, "connect[s] us with an enterprise originating in individual vanity—an enterprise got up for 'glory.'" According to John Savage of Tennessee, Arctic exploration generated lots of florid prose in Britain but little else. The British had entered the region "for the alleged purpose of making important discoveries in the northern part of this continent," declared Savage, "but, in reality, only to make a little material for royal flattery and newspaper glory," which "might for a time amuse the English public." The *Philadelphia Daily Ledger* pointedly brought the issue before its readers. "We have yet to learn that our government have any constitutional right, or are under any moral obligation, to hazard the lives of American officers and seamen, and appropriate the property of the American people, to such search." Even some Whigs remained skeptical. "Enterprises which spring from a desire of glory," warned William

FIGURE 3. Sir John Franklin. From Kane, *United States Grinnell Expedition,* frontispiece.

FIGURE 4. Henry Grinnell. From Kane, *Arctic Explorations,* vol. 2, frontispiece.

Seward of New York, "are very apt to end in disappointment." Despite their disagreements, Democrats joined Whigs to approve the resolution to nationalize Grinnell's voyage by a wide margin. America's first Arctic expedition sailed a few weeks later.[21]

Focused on rescue and cobbled together from public and private resources, the Grinnell expedition seemed to have little in common with earlier U.S. exploring expeditions. It did not seek trade, science, or geographical discovery, though its advocates hoped for such benefits. It was, by any standard, an exceptional enterprise, one shaped more by contingent events (the disappearance of Franklin and the munificence of Grinnell) than by a set of policies or national priorities. What significance, then, attaches itself to America's first foray into the Arctic, apart from the fact that it was first?

There are two answers. First, the Grinnell expedition represents the culmination of a shift in ideas about exploration that had emerged since the 1820s. Although inspired by a series of contingent events, the expedition was launched because Americans were prepared to judge it by a new set of values. It is hard to imagine the grimly practical Congresses of the early republic approving of an expedition that even its supporters admitted stemmed chiefly from the "love of glory." That the Congress of 1850 proved so willing to support the project shows us that it had learned some lessons from the European expeditions of discovery in general and the British search for the Northwest Passage in particular: that many of the greatest benefits of exploration were symbolic, that they accrued to national reputation rather than to commerce or territorial gain, and that however effervescent its objects, expeditions for glory were valuable enough to stake money and lives on their prosecution.

Second, the Grinnell expedition constituted a project well suited to the fractious political world of the 1850s. America's new attention to the Arctic offered a happy respite from the wrangling over American policy in the West. Two years earlier, the Mexican War had dramatically expanded the western territories. As Congress discussed the rescue mission, debate about whether these new lands would be free or slave still raged. The Missouri Compromise was falling to ruins, and some southern senators already threatened secession. Within this context the Arctic expedition offered a relatively safe means for northerners and southerners to express national pride. As the Arctic debates came to an end, the Ohioan Joseph Cable could still satirize the state of affairs by offering an amendment to the resolution: "*Provided*, That neither slavery nor involuntary servitude, except for crime, shall exist in any country or countries which may be discovered by said expedition, about the North Pole." Congressmen received the amendment "with much merriment." Before entering the Arctic, Americans were learning the lesson that the British Admiralty had learned years earlier: Arctic exploration constituted a safer form of conquest, offering many of the advantages of war without the messy commitments of empire.[22]

A Man of Science and Humanity

Elisha Kent Kane

WHEN Elisha Kent Kane died on 16 February 1857, he was one of the most celebrated men in America. He had come to the public's attention four years earlier when he led an expedition into the Arctic in pursuit of Sir John Franklin. Kane's ship *Advance* became trapped in the ice, preventing him from returning home as expected the following year. It came as a surprise when Kane appeared in the summer of 1855, two years after his departure, with most of his crew alive. Newspapers rejoiced in Kane's return, chronicling his party's escape from the high Arctic in small whaleboats and its passage down the treacherous Greenland coast. After his return, Kane (see fig. 5) grew progressively weaker from bouts of rheumatic fever. When he died in Cuba at the age of thirty-seven, Americans across the country mourned his loss. His extraordinary funeral cortège, which took three weeks and passed through six states, offers some measure of his popularity. After leaving Cuba, the cortège traveled up the Mississippi River before heading east by rail to Kane's final resting place in Philadelphia. Along the way, it made dozens of stops for local processions and memorials. In New Orleans, the mayor and a military company escorted his casket to City Hall, where he lay in state. In Cincinnati, his body was carried through crowded neighborhoods draped with banners where people were dressed in mourning. Baltimore's streets were thick with "ladies, who numbered thousands in the houses and on the sidewalks." Before burial, his body lay in state in Independence Hall in Philadelphia, where thousands paid their respects over the course of three days, including the governor of Pennsylvania, military officers, scientific elites, and members of Congress. Only Abraham Lincoln's funeral cortège,

FIGURE 5. Elisha Kent Kane. From Kane, *Arctic Explorations*, vol. 1, frontispiece.

which made its way across the country eight years later, would prove more spectacular.[1]

Kane's escape from the Arctic was an impressive feat, one that seems to explain the frenzy of public mourning at the time of his death. For more than a century, the funerary spectacle has given conviction to Kane's eulogists and biographers that there are deeper meanings to be found in his voyage. Not only did Kane give Americans the vicarious thrill of adventure, they argue, but he also embodied for them the highest qualities of manly character. But character alone cannot account for Kane's popularity. He failed to achieve any of the major goals of his expedition. He lost three men, his ship, and, for a time, the command of half of his crew. Even credit for Kane's escape from the Arctic had to be shared with local Eskimo tribes, without whose aid he and his party would have perished.[2]

If manly character does not fully explain Kane's celebrity, what does? A closer look at the funeral pageant shows that mourners praised Kane for other traits, too, specifically his abilities as a man of science. In dozens of local ceremonies, officers and members of scientific societies such as the American Philosophical Society, the Academy of Natural Sciences, and the Maryland Institute joined thousands of mourners in escorting Kane's hearse through city streets. Kane's eulogists exalted him as a "a genius in science," a man whose "contributions to science laid the whole world under obligation" and by whose death "science has been deprived of an ardent advocate." Yet scientific contributions cannot fully explain Kane's appeal, either. Kane achieved none of the scientific goals of his expedition. He had sailed north in hopes of exploring the polar sea, mapping the northern coast of Greenland, and collecting scientific data and specimens, yet he returned with little of this promised bounty. If Kane's popularity as a manly hero raises questions, then, they are matched by questions about his appeal as a man of science.[3]

The answers to both sets of questions are linked together. The traits that helped Kane convince his audience of his cultivation and good moral character also help persuade men of science of his proper temperament, training, and seriousness of purpose. Why did Kane's poor showing in the Arctic have so little effect on his extraordinary reputation back home? As we shall see, it was because he had already established himself as a man of character and science before *Advance* left port. Events in the Arctic may have sparked Kane's meteoric rise as a national celebrity, but it was a reputation kept aloft by actions he had already taken. Campaigning for his expedition as an author and lecturer, Kane had crafted a public image that appealed to many audiences. When he returned, these audiences chose to view Kane's actions favorably, even when they conflicted with other facts and testimonials.

Understanding Kane's appeal is important because it shaped events beyond his expedition. The relationship between Arctic explorers and scientists started with Kane's public campaign of 1851-1853. One reason the relationship lasted so long is that it began so auspiciously. Through a combination of charisma and erudition, Kane gained the trust of the scientific community and made Arctic subjects the focus of scientific debate. His efforts complemented the work of the Coast Survey, the Smithsonian Institution, and the Naval Observatory, which were building a bureaucratic framework to link the practices of science and exploration together. Kane appeared as these agencies faced a growing cultural divide that separated the scientific men who organized expeditions from the military men who commanded them. Successful in formalizing the links between exploration and science, these institutions looked to Kane—who moved very easily among different groups—to make the links personal.

In so doing Kane established powerful precedents. Explorers who followed him modeled their approach to the scientific community on his. They tried to re-create his lecture campaigns. They emulated his lofty persona as an Arctic hero. His campaign made a mark on the scientific community as well. Scholarly societies that had sponsored him basked in the light of his popularity. In the ensuing decades they tried to recapture the magic of his campaign by supporting new Arctic expeditions, sometimes on very flimsy grounds. For example, the American Geographical Society, which had come of age during Kane's campaign, became the unofficial headquarters for U.S. Arctic exploration for the next twenty years, promoting polar expeditions even when they drew attention away from more pressing geographical work at home.

The roots of Kane's appeal extended beyond scientists and explorers. He impressed elite societies as well as small-town lyceums, men as well as women, northerners as well as southerners. In a sense his eulogists were right: there were deeper meanings to his campaign, but they were hard to see without the distance of history. Growing sectional discord gave Kane, a man of many talents, almost perfect pitch in dealing with U.S. audiences in the 1850s. By combining the traits of manliness with scientific character, he demonstrated that there was still a common denominator of values that bound middle-class whites together, even as greater cultural and political forces were tearing them apart.

THE FIRST GRINNELL EXPEDITION

Kane did not fit the stereotype of the rugged explorer. At thirty, he was a small, soft-spoken man. An attack of rheumatic fever as a teenager had damaged the valves of his heart and left him in fragile health. When he graduated from the University of Pennsylvania Medical School in 1842 he had intended to pursue a private medical practice. But his father, Judge John Kane, feeling that a private practice would tax his son's health, had arranged an appointment for him in the Naval Medical Corps. Apparently Judge Kane assumed that the quiet duties of a peacetime naval surgeon and the open air of naval voyages would be better for Elisha's delicate constitution. But the younger Kane used the experience for different ends. In the 1840s he traveled as extensively as he could, voyaging to South America, China, and India. In the Philippines he climbed into the mouth of an active volcano. In Egypt he was wounded in a brawl with Bedouins. When the Mexican War broke out, Kane traveled south, was seriously wounded, and lapsed into a twelve-day coma. By 1850 his wounds had healed, but Kane had grown weary of the "miserable tediousness of small adventures." As the United States had stretched its wings as an imperial power, it had given Kane opportunities to stretch his own. Despite, or perhaps because of, his fragile health, he had found his calling in exploration. When he saw a notice about the Grinnell expedition in a local paper, he volunteered immediately and was soon chosen as the senior medical officer.[4]

Not all of Kane's attention was focused on the north. Before Grinnell's ships reached the Arctic, Kane had set to work writing his account of the expedition. By custom the expedition commander generally wrote the first narrative of the cruise. But Edwin De Haven, who led the ships *Rescue* and *Advance* in pursuit of Sir John Franklin, had not been interested in the task. He handed off the project to Kane. Fortunately for Kane, there was soon something to write about. In late August 1850, *Rescue* and *Advance* entered Lancaster Sound to find British ships also searching for Franklin. While all the vessels lay together at the mouth of Wellington Channel, a sailor from one of the British ships found three graves on Beechey Island (see fig. 6). The dead men had been members of the Franklin party. Nearby lay the detritus of an extensive camp: foundation stones for huts, a forge, a carpenter's shop, meat tins, even a pair of cashmere gloves laid on a rock to dry. Unable to find a written record, the explorers could only make guesses about the camp, the graves, and Franklin's course. Before De Haven could exploit the discovery, pack ice locked them in. They drifted with it northward up Wellington Channel and then south into Lancaster Sound. There *Rescue* and *Advance* slowly drifted east, reaching Baffin Bay by the beginning of 1851 and, once free of the ice, they reached New York in October.[5]

The Grinnell expedition had little to show for its efforts. The party had been present when the British made their discovery at Beechey Island. And, in the course of being dragged north by pack ice, it had identified some new lands at the top of Wellington Channel. Yet from these expeditionary scraps the press made a meal. When the squadron sailed into New York in the fall of 1851, reporters burnished the voyagers' modest accomplishments. Editorials heralded the expedition as a test of character, one that the Americans had easily passed. At times confidence gave way to cockiness. The *New-York Daily Times* wrote: "The main things wanted for such a service are zeal, intelligence, and perseverance; and these are the very qualities in which the Yankees have the advantage over other people. The men connected with the last expedition were equal to any in the world; in toughness and capacity of endurance,—in energy and willingness to endure

FIGURE 6. Graves on Beechey Island. Elisha Kent Kane's narrative of the first Grinnell expedition (1850–1851) made him famous to American readers. He commissioned James Hamilton, an established marine painter, to produce illustrations for the text, emphasizing the moments of highest drama, including the discovery of the graves of members of the Franklin party. From Kane, *United States Grinnell Expedition*, following p. 162.

fatigue and all kinds of privations for the sake of accomplishing their purpose."[6]

Gone were the critical debates about Arctic exploration that had swirled through Congress and the press in 1850. What explains the shift in press coverage? Documents say little. But they show one thing of interest: writers were eager to dress this modest expedition in the full regalia of nationhood. This, perhaps, partly explains the Grinnell party's cheery reception at home. When the expedition existed only as an idea, reporters and legislators freely debated its merits. Once under sail as a national enterprise, however, it operated according to a different set of rules. Critics closed ranks with supporters to urge the expedition on. Few were willing to publicly challenge its value. With critics muted and Franklin still missing, the press soon hummed with talk about a second U.S. rescue expedition.

Under normal circumstances, Kane would never have been a candidate for command of an expedition. He was, after all, a medical officer with little experience leading men. But Kane's detailed journal and status as the official expedition historian soon made him a favorite of the press, which sought him out for information about the cruise. It also brought him into closer contact with Grinnell, who saw Kane's growing popularity as a means of promoting a second voyage. Although Grinnell had the means to send his own expedition north, he believed that it was important for his ships to sail under naval command. The tight discipline of a naval ship, Grinnell believed, was essential in confronting the dangers of an Arctic voyage. As he viewed it, Kane's ability to captivate audiences would prove useful in persuading Congress to order the navy to carry out a second mission.

Yet it would be a difficult case to make. When the first expedition had left New York in 1850, John Franklin had been missing for five years. Now seven years had elapsed since Franklin's departure, and it would be another year before an expedition could leave for the Arctic. Finding the missing explorers alive seemed increasingly unlikely. Weighed against the slim hope of finding them were the considerable costs and risks of a new expedition. If the experienced Franklin and his company of 129 men could vanish into the Arctic, so, too, could an American party. Responsibility for any deaths would fall on the navy. And it would still have to pay for the salaries and provisions of the officers and crew. Both Grinnell and Kane knew that they had to make a convincing case that Franklin and his crew were alive and that they could bring them safely home.

THE OPEN POLAR SEA

In the theory of the "open polar sea" Kane and Grinnell found their justification for a new search for Franklin. Since the Renaissance, geographers had suggested that the high Arctic regions might be free of ice. The maps of Mercator represented the polar regions as open and watery channels at the top of the world. In the seventeenth century, mariners corroborated these maps by claiming to have voyaged across this polar sea on their way to Asia. Their travels were discussed in the *Transactions of the Royal Society*. In the late eighteenth century Russian explorers and whalers gave a name, "polynia," to such regions of open water, which they had frequently observed north of Siberia. Ferdinand Petrovich Wrangel observed a polynia in his passage through the Bering Strait. As British explorers revived their Arctic quest in the 1820s, they also reported seeing vast expanses of open water. These reports sometimes correlated open water with an increase in temperature and wildlife in the region. Although no one aboard De Haven's ships had seen this open polar sea directly, De Haven himself thought that he had found evidence of it at the top of Wellington Channel where *Rescue* and *Advance* had drifted with the pack ice. There, looking northwest, he observed that "a dark, misty looking cloud which hung over [the channel]...was indicative of much open water in that direction." He also noted evidence of a milder climate to the north. "As we entered Wellington Channel, the signs of animal life became more abundant." He pointed out in his report that a British expedition under Captain William Penny had passed into this area shortly afterward and "reported that he actually arrived on the borders of this open sea."[7]

Kane argued that Franklin may have sailed into this open sea and become stranded there. He laid out his arguments in a letter published in the *New York Tribune*. The graves at Beechey Island made clear that Franklin had camped there in 1846. Valuable articles left at the camp, such as the cashmere gloves, suggested that his party had left the camp in haste. Perhaps the ice in Wellington Channel had broken up suddenly, Kane ventured, and Franklin had abandoned his camp in order to take advantage of the open passage. Sledge tracks found to the north of the camp indicated that the party had moved northward up Wellington Channel. Observations from Kane's own drift up Wellington Channel in 1850 suggested that open water lay at the channel's northern mouth. Franklin's ships may have reached it and passed into the open polar sea. Once they had entered it, Wellington Channel may have once again filled with ice, preventing their escape. Although the Franklin party must have already exhausted their stores, they may have been able to survive on the marine life of this sea. The warmer climate and abundant wildlife observed by De Haven near the top of the channel suggested that food was plentiful. Kane's theory spread quickly in the popular press. Soon after it ran in the *Tribune,* the *New-York Daily Times* reprinted it. A few weeks later *Harper's New Monthly Magazine* brought Kane's arguments to an even larger readership: "It is the opinion of Dr. Kane that, on the breaking up of the ice, in the spring, Sir John passed northward with his ships through Wellington Channel, into the great polar basin."[8]

Kane's efforts to publicize his new expedition benefited from a growing demand for speakers on the national lecture circuit. Since the 1830s, public lectures had become a widespread form of education and entertainment in the United States. By the 1840s, more than thirty-five hundred U.S. communities had local societies that sponsored them. After Kane's discussion of Franklin appeared in the press, he started receiving dozens of requests to lecture about his experiences in the Arctic. That Kane started his speaking tour at a time when the nation was awash with lecturers might seem to diminish the importance of his campaign. But quite the opposite is true. As the lecture circuit became increasingly competitive in the 1850s, organizers sought speakers of certain interest and national reputation. Judging from their response to Kane, he met these requirements. Invitations poured in from New England and New York, where items about Kane frequently appeared in the press. But requests also arrived from other Middle Atlantic states, Ohio, Virginia, and Canada. Kane's appeal also reached across the social spectrum of self-improvers, from small-town lyceums and mechanics' institutes to well-to-do scientific organizations such as the American Philosophical Society. Although he attracted the attention of exclusively male organizations such as the YMCA, most organizers sought Kane for mixed audiences, assuming that he would appeal to men and women alike. Even exclusively female forums such as the Camden Young Ladies' Institute sought him out as a speaker. Grinnell was pleased. He saw Kane's lectures as the perfect opportunities to promote the new expedition. "My opinion is they [the lectures] would be well attended," he wrote Kane, "and do much good in enlisting the public's mind in favor of further efforts for the rescue of Sir John."[9]

But the opportunity offered by these lectures brought risks as well. The diversity of audiences that spanned the spectra of class, gender, education, and region made Kane's appeal more difficult. He might win over the working-class men of a mechanics' institute with stories of Arctic adventure, but would they impress the naturalists of a scientific society? Discussing scientific arcana of the Arctic would help his case with men of science but might prove boring to influential but less well-educated listeners. Kane solved this problem in part by narrowing his commitments. He turned down engagements at small or distant venues to focus on more influential scholarly organizations such as the American Philosophical Society, the Maryland Institute, and Grinnell's American Geographical Society (AGS). These organizations offered better opportunities to influence opinion makers. But their members did not share uniform backgrounds or educations. In the 1850s, the members of these groups came from very different backgrounds. Wealthy gentlemen with little formal education took their seats next to degreed men of science. And because many of these lectures were open to nonmembers, audiences tended to be even more diverse than the membership rolls, drawing wives and daughters, city residents, and even passers-by.[10]

Kane used the open polar sea theory to appeal to these different audiences. For individuals disposed to the heroic and patriotic aims of exploration, the existence of such a sea offered a plausible way to explain Franklin's where-abouts and continued survival. In this sense, the argument served the aims of a second rescue expedition. For scholarly audiences, the polar sea theory itself, and the mission to confirm it, touched at the heart of current scientific research. Kane's discussion of the open polar sea particularly interested Matthew Maury of the Naval Observatory and Alexander Dallas Bache of the Coast Survey. Both men had conducted research on ocean currents. They believed that the Atlantic Gulf Stream carried tropical waters far to the north of Europe. If Kane could confirm the existence of an open polar sea, it would corroborate their claims by suggesting that the Gulf Stream passed into the Arctic basin itself, keeping the waters there warm and free of ice. At stake was more than a confirmation of their theories about the Gulf Stream. Naturalists knew that the tropics received more energy from the sun than did the other parts of the globe, yet the tropics did not get warmer over time. If the equatorial waters traveled to the pole, then perhaps this was the way oceans maintained their relatively stable temperatures. The poles and the tropics must be tied together by a complex circulation system much like that of a living organism. Maury and Bache were both philosophically committed to viewing nature as a system of global, lawful, and interconnected phenomena, and proof of an open polar sea would confirm their basic beliefs about the world.[11]

SCIENCE AND CHARACTER

Kane's theory not only gave purpose to a new expedition but also showcased his strengths as an explorer. By using scientific arguments to promote his mission of mercy to find Franklin, Kane framed himself as a man of science as well as a man of character. These attributes became increasingly important as his campaign revealed him to be the face of the second expedition. As Grinnell became aware of Kane's growing power as a public figure, he urged him to use it to best effect in promoting the voyage: "We must get all out of the government we can in aid of *Kane's Arctic Expedition*. It is no more than just; if the people of the U. States could speak on the subject, they would say pay every dollar ... this new Expedition in the papers is called Grinnell's new Arctic expedition. I am not willing it should be so called. If there is any honour attached to it, it belongs to you. You must and shall have it. You are the heart, soul, and master spirit."[12]

It is not surprising that scientific sensibility and moral character served as the most visible attributes of Kane's "master spirit." Both qualities carried considerable weight among middle-class Americans at mid-century. They had become important means of social evaluation as other measures of class status such as wealth and lineage had begun to break down. Moreover, they fit in well with reform movements sweeping through white Protestant culture in the 1840s and 1850s. Science offered many Americans an important avenue for character-building because they considered the study of nature to be a morally uplifting activity. Within this context, Kane's accounts of his own arduous Arctic botanizing, however far removed from Americans' everyday experience, reinforced a idea with which they were well acquainted: that Kane's time outdoors had left him supremely self-improved.

The connection between science and self-improvement helps explain a riddle of Kane's campaign. For all that Kane discussed Arctic science in his lectures—and was widely acknowledged for it in the press—almost no one commented on his specific findings or their significance. Those who wrote about science in Kane's campaign did so to acknowledge his high character, not to explain his contribution to geographical or scientific knowledge. In other words, commentaries most often discussed his scientific work as a means by which to understand Kane himself rather than the Arctic regions. This helps explain why Americans who had never seen him lecture or knew any of the details of his work spoke so earnestly about his dedication to science. A Maine parson, for example, later recalled for his congregation, "Nothing perhaps strikes one more, than the mass and richness of [Kane's] scientific results, attained under such incredible hardships and disadvantage." At issue for the parson was not the meaning of the results themselves (which are never mentioned) but the fact that they "show a quality which we may fairly call the heroism of a trained and cultivated intellect."[3]

But interest in science and in Franklin does not fully explain Kane's appeal. Audiences responded not only to the subjects of his lectures but to the way in which he presented them. He combined his discussion of science and Arctic geography with intensely subjective passages about Arctic life. Such passages gave vibrancy to Kane's talks and allowed his audience to experience his self-improving struggles vicariously. "I am with the party in all their weary journeys," a women reader wrote, "and when I turn to gaze on the dark magnificent landscape, I can almost realize the solemn, the dreadful stillness of the Arctic night as [Kane] so vividly and pathetically describes." Another reader credited Kane with "transport[ing] us, by his graphic pen, into scenes we scarcely realize as belonging to the earth we inhabit." Men and women who attended Kane's lectures gave similar accounts of his ability to project them into the scenes he described. Recalling his address to the Boston

Lyceum, one man wrote about it in personal terms: "The brave explorer took us into winter quarters, where we were literally shut in by walls of ice, which perchance might become a living tomb, so real did it all seem, verily I felt my blood congeal with dread." As these passages illustrate, Kane had the power to direct the attention of his readers and listeners inward, to make them envision themselves as figures in a different world. They did not forget, however, the literary gift of their narrator, a style of writing that put him in the same company as other scholarly men of character such as Alexander von Humboldt and Ralph Waldo Emerson.[14]

For elite scientific audiences, Kane used his association of science with character to different ends. Specifically, he used his growing reputation as a man of character to quell doubts about his scientific abilities and forestall criticism about the poor results of the first expedition. Although he did not have special responsibilities as a science officer, members of the Washington scientific community looked to him because of his training and his role as senior medical officer to speak for its scientific work. When the Coast Survey began working up the data from the first expedition, the staff geodesist wrote to Kane asking him to explain missing and shoddy data. Kane admitted to him privately that the scientific work had been a debacle. Even straightforward measurements such as temperature had proved too much for the expedition party: "Our thermometers have never undergone comparisons either among themselves or with standard instruments at home. They have been mere pilgrims wandering about at the option of the observing officer—sometimes on the windward stanchions of the housings, sometimes on a pole a few yards from the brig—sometimes hanging to the bobstays forward and while the brig was moving, in the precarious shadow of the main-mast. If an instrument was broken no comparison was made with regard to its substitute. Often two instruments, one a spirit and the other a mercurial thermometer, were used indiscriminately in the same twenty four hour [period] without any connective record or even note of the fact."[15]

The poor results of the first expedition made it important for Kane to assure the scientific community of his competence in, and commitment to, scientific research. He used his private collection of Arctic specimens, as well as the status afforded him as a university-trained physician, as leverage to gain greater access to the elite scientific community. In the fall of 1851, Kane began donating specimens from his collection to scientific organizations. He gave the Academy of Natural Sciences of Philadelphia Eskimo skulls, a polar bear skin, rocks, and fossils. He gave the Smithsonian an Arctic fox and, in his lectures there, promised that more treasures would follow from a second voyage. Via his connections to Grinnell and the AGS, Kane became acquainted with members of the New York scientific community. He also took up correspondence with other luminaries such as Louis Agassiz and Alexander von Humboldt to ask their advice about the second expedition. His lectures at the AGS were well attended and widely covered in the press. He sought out and received membership in a number of scientific societies including the AGS, the Academy of Natural Sciences, and the Maryland Institute. At his request, these societies set up committees to help him plan the new expedition's scientific objectives. By participating in the planning, committees generally became advocates for the trip, and Kane used their imprimatur to defend his expedition's commitment to science.[16]

Men of science showed their support in many ways. The American Philosophical Society raised $500 toward the salary of an expedition astronomer. Joseph Henry, Alexander Dallas Bache, and Matthew Maury spoke in favor of Kane before the Senate Committee on Naval Affairs, which considered the expedition. The secretary of the navy put Kane under special orders for the purpose of leading a second expedition and allowed other naval officers to join it under special orders. The high level of scientific support did not persuade Congress to put the expedition under naval command. But it gave momentum to the planning. Grinnell decided to support the trip on his own even though he continued to have reservations about the conduct of a civilian crew. In early 1853 Kane started outfitting *Advance* for a voyage into Smith Sound, off the northwest coast of Greenland, and, he hoped, the open polar sea as well.[17]

THE KANE EXPEDITION

Kane's legions of supporters could do little to help him in the Arctic. The expedition started off well enough, sailing from New York in May 1853 and reaching Smith Sound by the end of the summer. But the limits of Kane's prior experience with the Arctic, which consisted of drifting through Lancaster Sound aboard a ship, left him ill-prepared for the difficulties of his new mission. When he sent some of his men westward over the rough ice of Smith Sound in the spring of 1854 to set up a cache of provisions, the men soon collapsed from exhaustion and frostbite. Two members of the party managed to make it back to *Advance*, and Kane led a team to bring back the rest of the party. Although he managed to bring all of the men back alive, four of the crew had to have toes or parts of their feet amputated. Two of these men died shortly thereafter from the ensuing infections. To make matters worse, lack of fresh food had brought the onset of scurvy among the crew. Only the assistance of a local Eskimo tribe at Etah prevented the party's demise. Despite occasional ill-treatment by Kane, the men and women of Etah provided him with fresh meat, which kept the scurvy at bay.[18]

By the winter of 1854–1855, survival replaced the expedition's earlier focuse on science and on Franklin. The summer of 1854 had offered the brightest moment of the expedition when two of Kane's men had returned to *Advance* reporting that they had reached the open polar sea (see fig. 7). On a long trek up the Greenland coast, the men had reached a point at which open water stretched north to the horizon. The news had helped cast off some of the gloom that had begun to descend on the party as supplies ran low and scurvy continued to afflict the crew. Yet the crew's high spirits had not lasted long. When the ice that trapped *Advance* in her harbor did not break up by the end of that summer, a number of men began questioning Kane's decision to stay with the ship through another winter. More than half of the crew made plans to leave the ship and head south rather than spend another winter in the ice. Kane disapproved of these plans but allowed the "withdrawal party" to attempt an escape in August. The withdrawal party made little progress before winter arrived, and low supplies forced them to return to *Advance*. Their return added to the tensions that strained Kane's command of his men. He forced the men of the withdrawal party to mess separately. Later he had to put down a desertion by two other crewmen, attempting to shoot one who sought to return to the ship. In the summer of 1855, Kane himself finally acknowledged the necessity of abandoning *Advance* and led his party south in whaleboats. After a harrowing journey during which one man died, Kane and his men dragged the whaleboats for miles over pack ice and into the dangerous waters of Melville Bay. Then they made their way southward five hundred miles to the fishing village of Upernavik and later met with an American relief expedition that had been sent by Congress to find them.[19]

When Kane had not returned in 1854, his patrons and admirers had grown worried. Calls for a federal rescue expedition had come from Grinnell and from various scientific societies. Yet it is testament to Kane's growing popularity that such calls also came from other groups as well. In New England the Northern Young Ladies' Seminaries voiced their support of a rescue mission. In Congress, southern Democrats who had previously opposed U.S. involvement in the Arctic now championed a search for Kane and his men. Writing to Kane in 1855, his brother commented on the breadth of this movement: "It is Fortune that has made you the child of *your whole country*, and not of

FIGURE 7. The shores of the open polar sea. Although no one had yet sailed its waters, Americans became increasingly convinced that it was warm and free of ice. One member of the Kane party reported finding open water on his farthest trek northward up Smith Sound. From Kane, *Arctic Explorations*, vol. 1, following p. 306.

any part of it; a fact to be remembered when you draw near New York." When Kane reached the city in October 1855, he received an enormous public welcome. "The report spread throughout the City with the rapidity of a scandal in a country town," reported the *New-York Daily Times;* "newsboys ran along the street with bright visions of front seats in the Bowery pit for the next week dancing before their eyes, and they shouted loud and long, for they knew they had an extra that would sell."[20]

And sell it did. The story of Kane's return was soon wired to other cities up and down the East Coast. He received high praise in the press, which viewed his success in surviving two Arctic winters and then leading an escape from the Arctic in whaleboats as confirmation of his status as a man of character. In the months after his return Kane received more than a hundred requests for lectures. The New York stage producer James Wallack set a writer to work crafting the events of the expedition into a play. Kane's return ranked so high as a national news story that it quickly became a benchmark by which the press evaluated other events. Attempting to describe the depth of popular obsession with the doctrines of free love, *Frank Leslie's Illustrated Newspaper* compared it with "the fall of Sevastopol and the return of Elisha Kent Kane." Amid the exaltation of Kane, few voices offered any kind of critical review of his actions. Forgotten or ignored were the failures of the expedition: Kane had not found Franklin, explored the polar sea, or mapped the northern coast of Greenland; he had lost three men, his ship, and most of his scientific collections.[21]

FLAWS OF CHARACTER

Yet Kane had not forgotten the failures of the expedition. He fretted about producing a narrative that would live up to his new reputation and would appeal to scientific and popular audiences alike. Science, heroism, and manly adventure would have to be tailored to make them suitable subjects for family as well as scholarly readers. Such broad commitments made the process more difficult, and at times Kane feared that making the narrative too popular would "destroy its permanency and injure me." As a compromise, he took most of his scientific observations out of the body of the final text but convinced his publisher to print them—sixty pages in length—as appendices.[22]

Dealing with the issue of character proved even more vexing. High conduct was, after all, at the heart of Kane's reputation. How could he write a detailed account of the expedition without revealing his own deficits as a commander? Kane chose to focus his attention on the party's battles with nature and avoided, whenever possible, those that raged below decks. Writing about the dangerous transit of Melville Bay, he focused on the harmony and bravery of the men of *Advance*. "I could not help being struck by the composed and manly demeanor of my comrades. . . . All—officers and men—worked alike." At the same time, he played down the numerous conflicts between him and his men. In the seven hundred pages of the two-volume narrative, he devotes four pages to the events of the secession, which he calls the "withdrawal." Absent is the fury Kane expressed in his private journal at the "deserters," whom he characterized as "little children" with "puerile opinions." In the published narrative, they emerge as "men confident in their purpose" and receive, on their abject return, "a brother's welcome."[23]

Kane's generosity of sprit did not extend to the Eskimos of Etah. These "strange children of the snow" appear in his narrative so bereft of civilization, so proximate to humanity's natural state, that *character* is a word that scarcely applies to them. Kane shows no anger in describing the Eskimos' acts of theft because they were "conducted with such a superb simplicity" that they were much like the wrongdoings of children. As such they could not be judged according to the standards of adult (that is, civilized) behavior. By the same analysis, their acts of bravery and kindness, which Kane does acknowledge, were also the products of savage naïveté. In this way he plays down the crucial actions of the Eskimos in keeping him and his crew alive. Denied character in a narrative that is built on it, the Eskimos form a background for Kane and his crew, whose actions play out almost unassisted.[24]

Why did Kane offer such a dismissive portrayal? Motives are difficult to attribute. By playing down the role of the Etah natives in the survival of the party, Kane could feature his own efforts. And by creating a chasm of difference between the Eskimos and the visitors, he made the gulfs of opinion within his party seem smaller by comparison. But Kane's views of the Eskimos were also shaped by cultural attitudes common to Americans of his era and background. The Eskimos were only the newest natives in a long history of New World encounters. In the vast sea of exploration literature authored by Anglo-Americans, one does not have to troll very long to find natives who act very much like Kane's Eskimos, who steal trinkets, lack modesty, and exhibit both kindness and curiosity. All of these traits—which bound savages and children together in a metaphorical union—had formed the boilerplate of exploration narratives for three hundred years.[25]

That being said, Kane's Eskimos are not generic savages. They bear the distinctive marks of his cultural heritage. They are descriptively linked not only to children but also to the archetypal savage for white Americans, the American Indian. For example, Kane "powwows" with the Etah Eskimos and describes one of them "as straight and graceful as an Iroquois." In their tribal priest Kane sees parallels to those "among our Indians of the West." And their rituals, he remarks, reminded him "through all their mummery, solemn and ludicrous at once, of the analogous ceremonies of our North American Indians." The Eskimos of *Arctic Explorations* thus became easily identifiable to most Americans as shorter, less threatening equivalents of the American Indians, and in so doing enter a framework of savagery well known to American readers.[26]

But all of Kane's work to praise his men and sideline the Eskimos failed to mend fences with his crew. As he labored on his narrative in 1856, he received word that other members of the party had started to write their own accounts of the expedition. Although custom and courtesy held that commanders published their accounts first, the men of the withdrawal party feared that his narrative would cast them as deserters. William Elder, who had started writing a biography of Kane, had already made unfavorable comments about the seceders in public. Angrily, one of Kane's officers wrote him to say that he found Elder's comments "discourteous to say the least and requiring correction." The most pressing threat, however, came from John Wilson, Kane's sailing master. Wilson had originally decided to leave with the withdrawal party and then changed his mind at the last minute. Kane had taken him back reluctantly. Angry about his treatment, and anxious to distinguish himself from the seceders, Wilson threatened to publish his own version of the withdrawal. In so doing he risked upsetting the delicate protections that Kane had put in place to preserve his own reputation and that of his company. Wilson's account of the withdrawal would undoubtedly raise questions about Kane's leadership. Although the popular press had heaped praise on Kane up to that point, it would be hard put to ignore criticisms coming from his highest-ranking officer. On the other hand, other members of the withdrawal party feared that Wilson, in his zeal to distinguish himself from them, would portray their actions as cowardly or mutinous. They, too, would have to publish accounts to protect their reputations.[27]

Faced with the prospect of competing accounts and a flood of individual grievances, Kane enlisted the help of his father to secure affidavits from loyal officers and crewmen. He hoped that such statements might be used as leverage against Wilson and the others. Whether Judge Kane secured the affidavits is unclear. Wilson continued planning to write an account in which the "humbuggery of that sly Fox [Kane] shall be exposed," but he seems never to have written it. Kane's narrative came out in the fall of 1856 and sold briskly. A year later, the volatile William Godfrey wrote his own account of the voyage, *Godfrey's Narrative of the Last Grinnell Arctic Exploring Expedition*, in which he criticized Kane and defended his own actions. But Godfrey's account caused little stir in the popular press. He was not an officer like Wilson or Hayes. Disliked by Kane and the seceders alike, he found no support for his claims. By the time Godfrey's narrative entered the marketplace, Kane's *Arctic Explorations* had already sold sixty-five thousand copies. Few could now compete against Kane's version of events.[28]

EULOGIZING KANE

Kane did not enjoy his celebrity for long. Flare-ups of rheumatic fever plagued him though the end of 1856, and he died in February 1857. Ironically, death completed the process that Kane initiated with his lectures and published works. It made him almost unassailable as a national hero. It also served as an occasion for Americans to explain and justify their noisy adulation. Many of these explanations came from eulogists who honored

Kane in local ceremonies held along the route of his funeral cortège. Others came from newspaper editorials trying to make sense of the spirited public mourning that followed Kane's procession like an infectious fever. Still others came from the pulpit and the fraternal lodge, where, months after his death, pastors, priests, and Masons reflected on the moral lessons of Kane's life and death. Even his crew offered their opinions about their commander's greatness, swallowing their bitter feelings for a while to take part in the gala of funerary events.

As we might expect, these eulogies are hardly balanced. They offer us little that is accurate about Kane's life. But they offer an excellent measure of the traits that, in the eyes of his supporters, made him a hero. As such they give us insight into ideals that Americans associated with Kane, particularly scientific sensibility and manly character. Because he had worked tirelessly to identify himself with these ideals during his campaigns, it is not surprising that they dominate his eulogies as well. But eulogists did not always follow in his footsteps. They gave voice to their own ideas, combining science and character in ways that were absent or latent in his work. Kane had mostly worked alone (or with Grinnell) to craft a persona that embodied the cultural values of his age. After his death, his admirers took up this project as their own. This is significant because it made Kane's place in American culture a collaborative project. If the picture of Kane that emerges bears little resemblance to the living man, it is a truer image of the man Americans wanted him to be.

It is significant that, in an era of rapid economic growth and specialization, eulogists did not honor Kane for one skill or accomplishment but rather for being a jack-of-all-trades. Given his careful appeal to popular and scientific audiences alike, perhaps it is not surprising that he found such praise for his versatility and well-roundedness. Eulogists repeatedly pointed to his integration of characteristics—moral, scientific, and artistic—as the hallmark of his greatness as an American. "We beheld him everywhere," the Reverend Charles Shields told mourners, "blending the enthusiasm of the scholar with the daring of the soldier and the research of the man of science." The Brotherhood of Masons praised Kane in similar terms. "His contributions to science laid the whole world under obligation; his writings embellish Literature; while his whole life is radiant with the divine spirit of Humanity."[29]

But the eulogists went further, crafting Kane into the image of the self-made man. In the process, they played down his formal education and links to the elite scientific community. William Elder showcased Kane's (dubious) resistance to traditional education in his hero-worshiping *Biography of Elisha Kent Kane*, published the following year. "Kane rebelled against the systematic learning imposed upon him by his childhood teachers," Elder wrote, "and was frequently reprimanded." Forced to discover nature without the aid of his teachers, Kane "tinkered at his own tuition in all the arts, sciences, and polytechnics of the boy-system of self-culture." At the University of Virginia and the University of Pennsylvania Medical School, Elder wrote, Kane acted as a scholarly iconoclast, pursuing only the knowledge that had "serviceableness to his actual uses." Elder thus found a way to disconnect Kane from his considerable education and add him to the ranks of other great "self-taught" men of science such as Humphry Davy and Benjamin Brodie. Even John Frazer, an esteemed professor of science at the University of Pennsylvania, acknowledged that Kane's multiple gifts had allowed him a means of transcending traditional scientific training. "His was a scientific mind... by which he was enabled to see the connection which lay between phenomena," Frazer told a Philadelphia congregation. "It was not merely in recording science that Dr. Kane excelled, but it was in that beautiful disposition which enabled him to see something beyond what is ordinarily considered science." Kane was no scholarly dilettante, he implied, but a man whose individualism and breadth of knowledge had made him a more capable observer of nature.[30]

Eulogists adopted Kane's focus on character but did so in a manner that might have given Kane pause: by accentuating his physical frailty. During his campaign to lead the second Grinnell expedition, Kane had concealed his rheumatic attacks and fragile health. The popular press had followed suit. But after his death Kane's illness became a means of accentuating the power of his character. His strength of character "raised his feeble frame above bodily weakness," a Philadelphia man declared, "and enabled him to triumph over cold and hunger." Kane's delicacy extended beyond his health. He had continually suffered seasickness "whenever there was breeze enough to create the slightest swell," an officer declared proudly. "In fact, I believe no man but Dr. Kane would have persevered in the voyage under the accumulated diseases from which he suffered at that time." "His palate was delicate," a crewman remembered. But as conditions in the Arctic worsened, "he accustomed himself to eat puppies and rats." Even Kane's appearance started to change to fit his new role. Earlier reports had described Kane as "a middle-sized man" who stood at least 5 feet 6 inches. Now Morton told audiences that he "did not exceed one hundred pounds." Kane became delicate to the point of effeminacy. A housemaid at Kane's medical school remembered him for "his pretty gentle manners, . . . his sweet young face, and lovely complexion like a girl's." Praising him for his the care of his sick men, one newspaper observed that he had shown "the gentle qualities of a woman."[31]

Rather than diminish his status as a man, the image of Kane that emerged served to strengthen it. Implicit within these eulogies is the assumption that manliness existed as an inner quality rather than as a manifestation of physical skill or strength. "When something was to be done which required nerve and manhood," Hayes recalled, "a sleeping power was aroused within him, which sent palpitating heart, puffed cheeks, rheumatic joints, and scurvy limbs hastily to cover." Even Kane's death, coming fifteen months after his return from the Arctic, offered a final demonstration of his manliness: "He literally *postponed* the fatal crisis," Morton told audiences, "until his duty was fulfilled." It was in the Arctic, Reverend J. H. Allen told his congregation, that "the seed of death ha[d] been planted."[32]

Kane's frailty allowed eulogists to represent him as a new kind of peaceful hero, an American icon that did not conform to the old military models of heroism. "This man, whose lifeless form is the object of such emotions and such pageantry," one man declared, "in his life had never distinguished himself . . . on the bloody battlefield as a warrior." But Kane *had* distinguished himself as a soldier in the Mexican War, a fact in which most eulogists seemed uninterested or of which they were unaware. They singled out not only Kane's nonmartial feats but also Americans' *recognition* of them as praiseworthy. "We have city after city pouring out by thousands," another eulogist exclaimed, "to meet . . . the remains of a man who never fought a battle. It speaks well for the taste and character of our people when we see such regard paid the disciples of science,—to honors won in the peaceful but laborious investigations into the earth's formation." In this way, eulogists framed Kane's greatness as a reflection of values already held by Americans.[33]

Kane's pacific heroism played well at a time when Americans' views of war had become increasingly ambiguous. The enthusiasm that had accompanied U.S. victory over Mexico in 1848 had faded. The new territories ceded to the United States had upset the balance of free and slave states. On March 6, 1857, as the train bearing Kane's body headed east toward Philadelphia, the Supreme Court rejected Dred Scott's appeal. War between abolitionists and pro-slavery settlers had broken out in parts of Kansas. Within this context, Kane's "war" against the Arctic remained a relatively safe outlet for national pride. Kane's war was also consistent with northern pacifist movements that had come into vogue in the 1840s and 1850s. Finally, Kane's Arctic war offered a story that could finally compete with real wars abroad. The American press had shown its envy of the British war in the Crimea that had so dominated world news. Thus when Kane had arrived home in 1855, the *New-York Daily Times* had delighted in comparing the bravery of Kane and his crew to the British in their battles with the Russians. "They had looked upon vast solitudes of eternal ice and snow, which human beings had never gazed on before, and had encountered hardships and dangers in the solemn darkness of two Arctic winters greater and more trying than the soldier faces in mounting the deadly breaches of a beleaguered citadel. The scene of public attention was changed from the Crimea to the North Pole."[34]

A year and a half later, eulogists were drawing the same parallels. "I do not pause to ask whose was the greater heroism," one man declared, "those who fought within and without Sevastopol, or those eighteen American men who, clustered in the little cabin of the *Advance*, watched and suffered during two Arctic winters. . . . Our Philadelphia hero was with the heroes of peace, in solitude, in silence and suffering." But the allusions to battle extended beyond the Crimea. One Boston minister compared Kane to the Duke of Wellington, who had conquered Napoleon at Waterloo. Like the great British commander, Kane, too, "had conquered, though not with marshaled army of foot and horse, but with the trained activities of a serene and lofty nature." Morton also used this comparison. "Dr. Kane overcame more terrific obstacles than Napoleon ever dreamed of, in a holier cause. . . . What was the passage of the Alps; what even the retreat from Moscow, compared to the horrors of two sunless winters?" Kane was also praised for his "Caesar-like gifts," and his expedition found itself represented as a "Christian Iliad." In this way, Kane's actions in the Arctic became a kind of noble warfare, which could be described in the language of war but transcended the violence of war for loftier goals.[35]

This noble war against the Arctic did not include Eskimos. In the same manner as Kane had, eulogists made the Eskimos noncombatants by representing them as savage children. Kane "carried our banner of peace," one speaker declared, "to the frozen children of the Pole." That he had become a friend of these poor savages, Elder declared, demonstrated his moral superiority to them. "The heart so tender and true to objects so repulsive as these could not be insensible to the charm that there is in childhood, in its beauty and innocence." By emphasizing character rather than strength, even the feats of the expedition's finest native hunter, Hans Hendrik, became minimized. Notwithstanding Hendrik's critical role as provider or his trek with

Morton to the open polar sea, he remained a child when compared to Kane. "Poor Hans he looked upon as his own personal charge," one crewman remembered, "and humored his whims and wishes as he might have done a child's." By presenting Eskimos as children, eulogists not only distanced them from the manly image of Kane but diminished the importance of their actions in saving Kane and his party.[36]

CONCLUSION

More than anyone else, Kane made Arctic exploration an American enterprise. Although U.S. interest in the Arctic had sprouted from British roots, he grafted different practices to polar exploration that allowed it to thrive in the United States. Whereas most British expeditions to the Arctic had been top-down projects, planned and funded through the Admiralty, the Kane expedition had grown up as a private initiative between Kane and his patron, Grinnell. Both men had hoped that the popular campaign would induce Congress to place the expedition under naval command. But while the U.S. government had become sensitive to the symbolic value of discovery expeditions, it shied away from the Arctic as a stage too perilous for sustained federal exploration. Washington's reluctance forced Kane to reconceive his expedition. His popular campaign—originally intended to sway members of Congress—took on a life of its own, providing him with a means of cementing support among various audiences. As such, the process hinged on personal attributes: Kane's faculties as a writer and orator, as well as his abilities to connect with people of different backgrounds.

Yet Kane's personal qualities made such an impression only because they were so well suited to the mercurial conditions of American culture in the 1850s, a time of new entertainments, interests, and social controversies. Kane entered the homes of so many Americans because a new industry of newspaper, magazine, and literary publishing houses saw opportunities in his story and captivating prose. His stories reached new ears because a national circuit had grown up to exploit public lectures as a form of civic entertainment. And his reputation as a man of science blossomed because of the support of new scientific and geographical societies that came of age in the midst of his campaign. Thus, American expeditions to the Arctic took shape not only because of Kane's vibrant personality but also the way it meshed with new establishments that dotted the American cultural landscape. Not surprisingly, the success of his campaign ensured that other explorers would emulate it. That they repeatedly failed to do so reflects less on their personal shortcomings than on the changeable cultural climate, for which their campaigns proved ill-suited.

One result of Kane's campaign was to focus greater attention on explorers than on the expeditions they commanded. This contrasted with popular accounts of earlier expeditions to the Western Territories, which tended to focus on the actions of the expedition parties rather than their leaders.[37] Although the press and the public had praised the actions of the first Grinnell expedition, few writers singled out its commander, Edwin De Haven, for special merit. By comparison, Kane's popular campaign established him as the clear protagonist of his expedition, reducing the crew, the Eskimos, and even Grinnell to the roles of supporting characters. As a result, the press and the public were predisposed to view Kane as the symbolic point of focus for the voyage. In written accounts, it was Kane, not the expedition party, that embodied the traits of national character.

That Kane's image continued to shimmer after his death helped maintain this new focus. In 1858 William Elder's glowing biography of Kane appeared, only to be followed by a flurry of expedition histories by other publishers in the following decade. Traveling panoramas toured the country, recounting the explorer's life and Arctic adventures for paying audiences. The Masons renamed their New York lodge the Kane Lodge in his memory. Kane's girlfriend published an account of their relationship, *The Love Letters of Dr. Kane*. Mediums reported seeing him (sometimes accompanied by his spirit-world companion John Franklin), and Romantics such as Henry David Thoreau, Ralph Waldo Emerson, Walt Whitman, and Emily Dickinson wove Kane and the Arctic quest into their writings.[38] Perhaps most indelible of all was Kane's two-volume *Arctic Explorations*, which went through multiple editions and sold more than 150,000 copies in the following decades. If it served as a source of inspiration for readers, it also operated as a guidebook for future explorers, who used it to emulate Kane's persona and popular campaigns in the decades to come.[39]

———— ✳ ————

CHAPTER THREE

An Arctic Divided

Isaac Hayes and Charles Hall

IN 1854, as Elisha Kane's party searched for Sir John Franklin in the high Arctic, a British explorer far to the south found the first real evidence of Franklin's whereabouts. On the northern edge of the American continent, Dr. John Rae of the Hudson's Bay Company met two Eskimo men from the Pelly Bay tribe. One of the men, In-nook-poo-zhee-jook, told him a strange story. He reported that at least thirty-five white men had starved to death on a distant river. Although In-nook-poo-zhee-jook had not been there himself, he showed Rae a gold cap-band reportedly found at the site. Rae bought the cap-band, and in the weeks that followed he bought additional relics from other Eskimos, including plates and crested silverware, that identified the dead men as members of the Franklin expedition. That summer, Rae returned to England to report the news to the British Admiralty. He carried other disturbing news reported to him by the Eskimos: "From the mutilated state of many of the corpses, and the contents of the kettles," Rae wrote, "it is evident that our miserable countrymen had been driven to the last resource—cannibalism—as a means of prolonging life."[1]

Franklin — cannibalism

The news touched off a storm of controversy in Britain. In the nine years since Franklin's departure, the popular press had transformed him and his party into national heroes. Some still held out hope that the missing explorers remained alive. This faith flowed in part from the belief that Franklin and his men possessed the highest character of the British people, character that would aid them in their struggle to survive. "It seems scarcely possible," the *London Examiner* had argued in 1850, "that the whole hundred and twenty-six men, the flower of our navy, can have sunk entirely hopeless under their difficulties, and perished already." Rae's report suggested that the Franklin party had not only lost its war against the Arctic but had been reduced to the behavior of savages in the process. The British press and public accepted the news reluctantly. Critics attacked Rae's reliance on Eskimo testimony as the weak link in his report. The testimony of savages, they argued, had to be weighed against the far more reliable attributes of British character. It seemed to them that, faced with starvation, Franklin's men would have died rather than resort to cannibalism. Charles Dickens championed this position in his magazine, *Household Words*. "The noble conduct and example of such men, and of their own great leader himself," Dickens wrote, "outweighs by the weight of the whole universe the chatter of a gross handful of uncivilized people, with a domesticity of blood and blubber." Rae rebutted Dickens in turn, and their heated exchange continued through the winter in London papers and magazines.[2]

In the United States, the Rae-Dickens debate played out more quietly, but it had a profound impact on the course of American exploration of the Arctic. Whatever Americans may have concluded from the Eskimo reports, Rae's relics offered persuasive evidence that Franklin's expedition had traveled *south* from Lancaster Sound and not *north* into the high Arctic, as Kane had suggested. By placing Franklin's men on or near the American continent, it exploded the link that Kane had so carefully forged between Franklin and the open polar sea. In effect, Rae's report ensured that the two most compelling goals of Arctic exploration, the search for Franklin and the search for the great polar ocean, should operate as geographically separate activities. In so doing, it forced explorers to choose between missions of rescue and science. The choice affected the way explorers ran their campaigns back home, compelling them to tailor different approaches for scientific and popular audiences. At times, explorers courted one group at the expense of the other. If Kane fostered an easy alliance between scientists and Arctic explorers, Rae's discovery threw this alliance into doubt.[3]

These tensions become evident in the campaigns of Isaac Hayes and Charles Hall, who aspired to lead Arctic expeditions in the late 1850s (see figs. 8 and 9, pp. 57 and 67). Both men struggled to adapt to the new conditions generated by Rae's report. In linking the discovery of the open polar sea to the search for Franklin, Kane had been able to promote his expedition as a mission of "science and humanity." By contrast, Hayes and Hall had to choose either science or humanity as the principle focus of their expeditions. Scientific audiences tended to support Hayes's expeditions to the polar sea, whereas lay audiences preferred the human interest of Hall's searches for

FIGURE 8. Isaac Hayes. From Hayes, *Open Polar Sea,* frontispiece.

Franklin. Despite their best efforts, their campaigns narrowed to attract the interests of these different groups, making it especially difficult for either explorer to dominate Arctic discourse and generate the broad-scale enthusiasm needed to promote and bankroll their expeditions.[4]

In order to attract their respective audiences, Hayes and Hall chose to present themselves in different ways. Hayes fashioned himself in the image of Kane, as a man of science and cultivation. Hall took his inspiration from Rae and chose to live with and learn from the polar Eskimos. Both strategies had advantages and disadvantages. The scientific rhetoric of Hayes's campaign made him the darling of the scientific community, but it left popular audiences cold. Hall's portrayal of Eskimo life thrilled the public, but it also made it difficult for him to represent himself as a cultivated man, a man who embodied the highest values of American civilization. By building his reputation as an explorer on his faith in the Eskimos and his ability to live like them, Hall eroded the bulwark built by Kane and others between savage and civilized ways of life. Thus, the rivalry between Hayes and Hall had implications that extended beyond their campaigns. It laid bare a fault line between the scientific and popular audiences that would widen over time, presenting grave challenges to explorers in the decades ahead.

ISAAC HAYES

Isaac Hayes's experiences as a member of the Kane expedition of 1853 shaped his own Arctic campaign in many ways. Serving as Kane's surgeon, he had learned at first hand about the terrible hardships of Arctic exploration. In addition to treating the injuries and illnesses of Kane's party, Hayes endured scurvy and lost several of his toes to frostbite. And having safely returned, he learned a second important lesson: the Arctic held a fantastic power to captivate the American public. In 1855 he participated in the jubilant homecoming of *Advance* and saw the way Kane's glory had eclipsed that of all other members of the expedition. Although Kane played down his criticism of Hayes and other members of the withdrawal party in his narrative, Hayes and the others felt left out of the spotlight and concerned that Kane's private rebukes had poisoned their reputation among members of the exploration community. In 1857, when Hayes announced his plans to lead an expedition, he justified his trip as the completion of Kane's work. He made the open polar sea the geographical focus of his mission, just as Kane had done six years earlier. He also adopted Kane's route up Smith Sound as his gateway to the high Arctic. Hayes even used Kane's strategy for raising support at home by combining private appeals to wealthy patrons with lectures, articles, and interviews that reached popular audiences. In effect, Hayes decided that the best way to lead his own expedition and redeem his reputation was not to renounce Kane but rather to present himself to his audiences as Kane's successor.[5]

Toward this end, Hayes presented himself as an explorer of scientific and cultural refinement. It helped that the two men shared the same scientific background; both graduated from the University of Pennsylvania Medical School and gained their Arctic experience as ship's surgeons. And Hayes used his experience and pedigree to reach the same members of the elite scientific community that had welcomed Kane in 1852 and 1853. Alexander Bache, Joseph Henry, and Matthew Maury had eagerly supported Kane in return for his commitment to Arctic research. Aware of their interests in global models of hydrography and meteorology, Kane had promoted Arctic research as an essential piece of the broader puzzle of terrestrial science. Because he had been forced to abandon *Advance* in the Arctic, he lost his chance to bring back the promised bounty of specimens and data. "The harvest which Dr. Kane so successfully reaped," Bache wrote, "is not all reaped." Similarly, Kane had brought back information about Arctic weather, but it was not comprehensive enough to render a clear picture of the polar winds. Such a picture was necessary, Henry argued, because it "is connected with the theory of the entire circulation of the atmosphere of the Globe." Kane had thus primed the pump for Hayes's expedition by raising scientists' interest in, and expectations of, Arctic research. By framing himself as a scientific explorer after the model of Kane, Hayes offered the elite scientific community a second chance to obtain the information that seemed vital to broader theories of terrestrial science.[6]

Winning the support of this small cadre of Washington elites also aided Hayes in wooing the broader scientific community. To do this, he left Washington to visit New York and the heartland of Arctic patronage, the American Geographical Society (AGS). With Henry Grinnell's encouragement, the AGS established a committee to help raise funds for the expedition. Hayes then headed to Baltimore, where he presented his plans before the American Association for the Advancement of Science (AAAS). Bache and Henry both served on the committee that was set up to assist him. Other powerful AAAS members, such as James Dwight Dana, served on the committee as well. Arctic science "demand[ed] further research," they concluded, and they justified the mission of the Hayes expedition to explore the open polar sea.[7]

Securing the support of select elite men of science helped Hayes gain the support of other scientific societies as well. Hayes's speech before the AAAS was published by the prominent *American Journal of Arts and Sciences*, which Dana edited. Lists of the prominent men who made up the AAAS expedition committee appeared in Hayes's correspondence with other scientific societies. Letters of support from Bache and Henry followed Hayes to the AGS, where they were read as introductions to his speech. (After the speech the audience learned that Bache and Henry had also pledged $1,000 in agency funds for the expedition.) The backing of such powerful scientists assisted Hayes in gaining declarations of support for his mission from the Academy of Natural Sciences of Philadelphia, the Lyceum of Natural History of New York, and the Boston Society of Natural History.[8]

But Bache and Henry could not iron out all the bumps of Hayes's campaign. Reflecting on it later, he called his Boston meeting with the American Academy of Arts and Sciences the "severest ordeal which I had yet passed." A gathering of fifty members grilled him about the details of his mission. Louis Agassiz, the celebrated Harvard University professor of zoology, questioned Hayes about his objectives for physical research and the collection of natural history specimens. But after this rough start, Agassiz warmed to Hayes and spoke at length about the value of Arctic exploration. At the explorer's request, the committee passed a "resolution of sympathy" for his mission. They even convened a public meeting at which Agassiz repeated his praise for the benefit of a wider audience. With Agassiz's name added next to those of Henry, Dana, and Bache among his public supporters, Hayes attracted more than four hundred subscribers for his expedition, most of whom belonged to scientific societies. Encouraged by the support of scientific elites and enthusiasts alike, Hayes began planning his departure for the summer of 1860.[9]

He also tried to drum up popular interest in his expedition. In Boston and Albany he set up "citizens' committees" to raise funds for the expedition. He targeted appeals to Masonic lodges that had supported Kane. He also sought out contacts among potential patrons, giving lectures to the New York Chamber of Commerce and the Board of Trade in Philadelphia. In 1860 he published an account of the withdrawal party, An Arctic Boat Journey, hoping that it would raise money for his expedition. These efforts did not bear fruit. Hayes's campaign did not generate the enthusiasm that Kane's had in 1852–1853. Nor did reviewers' praise of An Arctic Boat Journey for its "manly unaffected phrase" and "simple style" do much to sell copies of the book.[10]

Hayes had reproduced many of the elements of Kane's campaign—his Arctic route, scientific persona, and search for popular support— but he could not find a suitable substitute for Franklin. For fifteen years this British explorer had been the impetus for all of the British and American expeditions that had combed the Arctic. On both sides of the Atlantic, the search for Franklin had been invested with all of the elements of a good nineteenth-century story: mortal peril, moral urgency, and imperial contest. (One imagines that Americans especially savored the idea of Yankee volunteers saving an elite British squadron.) But for popular audiences who came to view the Arctic as a moral landscape and rescue as the bread and butter of an Arctic campaign, Hayes's scientific mission to the open polar sea seemed insubstantial.[11]

Concerned about Hayes's failure to ignite enthusiasm among the public, his scientific supporters attempted to boost his campaign with appeals to character. They described his mission using the same religious and military metaphors that had proved so effective for Kane's. In seeking the open polar sea, one supporter claimed, Hayes had embarked on an "Arctic pilgrimage." The search for Franklin did not constitute the only noble mission in the Arctic, another argued: "Science has its martyrs as well as religion." Indeed,

Hayes's pilgrimage rivaled the noble calling of the Franklin search. "Dr. Hayes goes in search of living knowledge," one supporter declared, "not in search of the dead." But his supporters also made clear that these lofty goals did not make Hayes's mission any less dangerous, thrilling, or manly. Hayes "is a soldier who asks for arms that he may fight for us." declared one AGS member. "If Providence . . . has decreed that you shall sleep as soldiers sleep on their own battle-fields, wrapped in the white Arctic sheet . . . your name [will] be worthily remembered as one who gallantly fell in a noble contest, truly on a field of honor." Hayes's scientific assault on nature put him in the company of his late commander, Elisha Kent Kane, who had once been exalted for his dreamy, gothic rendition of nature but who now found praise for "wrestling with Nature herself in her own fortified home" and for being a "Viking of science."[12]

THE HAYES EXPEDITION OF 1860

Inflated by such prose, Hayes's departure swelled with symbolic pretensions. He scheduled his ship's departure for Independence Day, even though at that late date he risked missing the short Arctic summer. When Bostonians gathered on Gray's Wharf, they viewed the schooner *Spring Hill*, now weighted under its new name, *United States*. A large crowd of men and a "goodly number of ladies" listened to the governor of Massachusetts and the president of Harvard University praise Hayes as a symbol of American pluck and zeal. Agassiz also addressed the crowds, claiming that the expedition served as evidence of America's ascent up the ladder of civilization. Navigation of unknown regions, he declared, "was the measure of civilization. . . . We had now reached a period when the merchants of this country were willing to dispatch a vessel for the sole purpose of bringing back knowledge instead of gold." But events resisted the symbols placed on them. The speakers hailed Hayes under a blanket of rain. And *United States*, cheered out of the docks as the denizen of "unknown regions," had difficulty sailing away from the coast. It dropped anchor for two more days to wait out unfavorable winds.[13]

The fumbled departure proved fitting. The Arctic did not cooperate with Hayes's plans, nor did it serve to forge stronger ties among his expedition, science, and the nation. Not all of the blame rested with the Arctic. Before reaching high latitudes, Hayes's party was showing signs of distress. In Greenland, the ship's carpenter died in his sleep. Meanwhile, two local Danish naturalists accused William Longshaw, the expedition's surgeon, of stealing their books and natural history specimens. A search of Longshaw's trunk turned up some of the missing items. With the Danish community in an uproar, Hayes quietly sent Longshaw home, where the surgeon told surprised reporters that he had returned because of snow blindness. But this did not silence talk about Longshaw's actions in Greenland. "This surgeon's rascality," Grinnell fumed, "had spread the whole length of the Greenland coast." And it would soon spread further. By the spring of 1861, Grinnell would learn the full story of the scandal from his son, who reported from England that the matter had become a topic of conversation among British explorers.[14]

Poor progress in Smith Sound further impeded the scientific goals of the expedition. Because Hayes had departed Boston so late in the season, he reached Smith Sound as summer drew to a close and pack ice began to fill the channel. When fierce gales churned the ice and battered *United States*, Hayes made winter quarters only eight miles north of the mouth of the sound. Their southerly position (south of Kane's 1853 quarters) guaranteed a long and arduous journey to reach the polar sea. Hayes made the best of a bad situation, ordering his astronomer, August Sonntag, to set up an observatory over the winter. He also set to work on building a natural history collection. With the promise of extra grog, Hayes's crew roamed the surrounding coast digging up Eskimo graves and gathering the remains for his ethnological collection. When one of the hired Eskimos, Merkut, stumbled across the bodies of two of her relatives aboard the ship, however, she protested bitterly to Hayes, who reburied the bodies. But ethics did not long hinder Hayes's collection of human remains. Merkut's husband Hans (who had accompanied Kane as well) brought Hayes the skulls of two Eskimos after ascertaining that they were unrelated to him or his wife. Other problems proved more difficult to manage. During the winter, Sonntag fell through the ice and died shortly thereafter of hypothermia, leaving the expedition without a science officer. Longshaw's departure had hobbled the scientific mission; Sonntag's death spelled its demise. No other member of the party could effectively conduct the astronomical, meteorological, and geomagnetic observations that Hayes had promised to the scientific elite back home.[15]

Hayes's trek to the polar sea gave him the only success of his expedition. In the spring of 1861, Hayes commanded a large sledge party as it crossed Smith Sound. The bitter cold, poor weather, and mountainous hummocks of ice slowed the party, and the crossing took a month. Turning all of the party back except one other man, Hayes pressed on along the coast of Grinnell Land (Ellesmere Island) until he reached a point where the pack ice seemed to disintegrate. The cracks in the ice beneath him, Hayes wrote later, "expanded as the delta of some mighty river discharging into the ocean." The sound seemed to be almost free of ice further to the north. Satisfied that he had found Kane's open polar sea and achieved a new record of "farthest north," he returned to the ship. Although Hayes had glimpsed the object of his polar desire, the long sledge trip had made it impossible for him to continue his expedition much longer. The journey had depleted his crew and used up his sledge dogs. Moreover, inspection of *United States* showed it to be severely damaged. With little chance of reaching the open polar sea on a second attempt, Hayes set sail for home, a year earlier than planned.[16]

The short voyage and meager results probably would have disappointed scientific and popular audiences under the best of conditions, but events back home had made a warm reception impossible. Months after his departure, the Civil War had consumed the country. Interest in the polar expedition, already tenuous in the popular press, evaporated as Americans became preoccupied by the battles of Harper's Ferry and Bull Run and events much closer to home. In contrast to the impressive crowds that had sent the party off a year earlier, no one turned out to cheer the arrival of *United States*. As it arrived at Long Wharf in Boston, Hayes left the ship and walked into town. "I felt like a stranger in a strange land," he wrote, "and yet every object which I passed was familiar. Friends, country, everything seemed swallowed up in some vast calamity, and, doubtful and irresolute, I turned back sad and dejected, and found my way on board again." The war had broken the nation's Arctic fever. Now, the *New-York Times* observed, the public viewed such exploits with the desultory detachment of "Lotus-eaters" who had received news of Hayes's return with a "languid half interest."[17]

The colossal presence of the Civil War in American life weakened interest in many peacetime projects. But it proved especially crippling to Arctic exploration. This was due, in part, to its stated objects—science and rescue—which seemed superfluous to Americans in time of war. But it also suffered because its symbolic appeal as a battle against nature competed too closely with war itself. Lest we forget, the British Admiralty had dusted off Arctic exploration in 1818 because it had potential as a *substitute* for war. Absent hostilities on the Continent, the Arctic had provided Great Britain with the theater for an ersatz war, which was fought by navy officers and crew and championed by officials, reporters, and writers in a flurry of patriotic language. As Americans gained interest in British Arctic expeditions, they learned to appropriate such language and put it to use in describing their own. Thus, Kane had used military metaphors, as had the British, not only to animate his expedition stories but also to frame them as the moral and national equivalents of war. Hayes had used the metaphors of war in *An Arctic Boat Journey*, in which distant glaciers boomed like artillery, and he reported Kane's musing on "how to look our enemy in the face." Indeed, critics had praised Hayes's narrative for its stories of warlike toil and torment. The value of the narrative was "not due to any signal success achieved by Dr. Hayes and his companions," one critic concluded, "but to a well-written narrative of terrible suffering, borne with a fortitude and patient endurance." But now Hayes's metaphors paled against the real horrors of battle. As Union and Confederate soldiers died in abundance on the fields of Virginia, the public took little notice of Arctic "Vikings," "martyrs," or "pilgrims." War erased the need for warlike heroes fighting campaigns against nature. And it gave the press a new standard of manly suffering against which Hayes's expedition fell short. His poor results, early return, and healthy crew flew in the face of his reports of hardship. Even as Hayes protested that damage to *United States* had forced his early return, reporters visiting the ship told a different story. "The vessel," one wrote, "looks better than she did when she sailed."[18]

Hayes fared little better in his reception among scientific audiences. The support of these societies became critical to Hayes as popular interest waned. They had largely funded his expedition, and now, as debts mounted, he looked to them again for help. Grinnell helped Hayes set up a lecture tour on which he would visit a number of prominent scientific societies in New York and Philadelphia. As the manager of Hayes's expedition funds, Grinnell had become increasingly worried that Hayes's debts could damage his own reputation as a man of science and character. Through the spring of 1861, Grinnell had tried to keep a lid on the scandal caused by Longshaw's theft. As word of the theft had spread, Grinnell had considered sending the Danish naturalists money in recompense for their stolen specimens. Because expedition funds were now too low to cover the crew's wages, he worried that if reporters discovered the diversion of money, they would interpret it as an attempt to cover up the scandal. He had hoped that Hayes would return with sensational discoveries that could be used to raise money and turn attention away from the Longshaw matter. "I think there would be no difficulties whatever," he confided to a friend, "particularly if the Dr was successful in making some discoveries and confirming Kane's open sea." In lectures before a number of scientific societies, Hayes put the best face on his voyage, offering twelve separate accomplishments in the fields of geology, geomagnetism, hydrography, and natural history.[19]

Yet his lectures did not significantly improve his popularity among scientific audiences. In fact, his debts threatened to alienate one of his most powerful scientific allies. In order to pay his crew's wages, Hayes took out a loan, using as collateral one thousand dollars' worth of scientific instruments borrowed from the Coast Survey. When he did not return the instruments, Bache wrote to him inquiring about them. Hayes was embarrassed to admit that he had used the instruments as collateral and did not have the money to buy them back. Even the popular press had begun to take a dim view of Hayes's scientific performance. "The result, as far as regards the main object, is absolutely *nil*," wrote the *New-York Daily Times;* "the scientific corps was altogether far too weak to authorize any great expectations." Having experienced at first hand the wild enthusiasm of Kane's reception, Hayes could only be disappointed at his own. "The expedition has been very coldly received here," he wrote Grinnell. "Everything is as bad as bad can be. . . . I have not yet seen anybody that gave me a welcome hearty and sincere." Discouraged, he stopped lecturing and took an appointment as a surgeon in the Union Army for the duration of the war.[20]

After the war was over, Hayes tried again to win over the public by publishing a popular account of his voyage, *The Open Polar Sea*. Although his account of the Kane expedition, *An Arctic Boat Journey*, had gained praise for its simple and unaffected style, *Open Polar Sea* was melodramatic and highly sentimental. What explains Hayes's shift in style? Perhaps he hoped that the new book would match the success of Kane's *Arctic Explorations*, a narrative praised for its intense, subjective passages. Kane's two-volume work continued to sell well and had become a benchmark of travel writing since its publication seven years earlier.[21] But Hayes did not have Kane's ear for language. Describing the quiet of the Arctic winter, Hayes writes: "There is no cry of bird to enliven the scene; no tree, among whose branches the winds can sigh and moan. The pulsations of my own heart are alone heard in the great void; and as the blood courses through the sensitive organization of the ear, I am oppressed as with discordant sounds. Silence has ceased to be negative. It has become endowed with positive attributes. I seem to hear and see and feel it. It stands forth as a frightful spectre, filling the mind with the overpowering consciousness of universal death,—proclaiming the end of all things, and heralding the everlasting future. Its presence is unendurable."[22]

Hayes's narrative did not ignite readers or critics as had Kane's. Even when reviewers praised Hayes's account, they criticized its style as mawkish and clichéd. After all, thousands of readers had already seen the Arctic described in this manner before. *Littell's Living Age* wrote that Kane's thrilling accounts had "somewhat dulled the edge of curiosity with which [other narratives] were formerly received by the public." The *Atlantic Monthly* agreed: "The sameness of the Polar world and Polar life wearies a little." But the literary climate into which Hayes's narrative made its debut had also changed. The grim realities of war had reduced the appeal of his dreamy, effete prose. His determination to maintain civilized standards in the Arctic clashed with the breakdown of civilization back home. When Americans were suffering through the battles and economic hardships of war, Hayes's readers learned that he had been making Arctic life more hospitable. Describing a birthday party for one of his men, he wrote: "There was a capital soup—*jardinière*—nicely flavored, a boiled salmon wrapped in the daintiest of napkins, a roast haunch of venison weighing thirty pounds, and a brace of roast eider-ducks, with currant-jelly and apple-sauce, and a good variety of fresh vegetables; and after this a huge plum-pudding imported from Boston." Such anecdotes seemed out of place in a country coming to terms with the death of more than half a million men.[23]

Whatever its shortcomings vis-à-vis the broader reading public, Hayes's narrative helped him reestablish his authority within scientific circles, principally through his discussion of the open polar sea. Although Hayes gave only gave two pages of the book to a description of the sea itself, he devoted a chapter to the evidence and arguments supporting the sea's existence. In the years following its publication, he capitalized on his book by authoring a number of articles and lectures about the sea's geography. His writings put him at the center of a small but active debate about Arctic hydrography. Although all of the participants in this debate accepted the existence of open water at the North Pole, they differed concerning its causes. Most attributed it to some heat source that kept the sea from freezing. Hayes defended Matthew Maury's theory that the waters of the Gulf Stream kept the sea open. Others attributed the open water to warm tropical winds or the magnification of the sun's rays at the poles. Still others credited the open water to causes unrelated to heat, such as the effects of the earth's rotation or the role of dense water in impeding the formation of pack ice.[24]

As Hayes considered organizing a second expedition, his participation in the debates about the open polar sea heightened his visibility within the scientific community. Such visibility not only helped him build the broad coalition of scientific and private patrons needed to launch another expedition; it also allowed him to shape the discourse about Arctic geography when it threatened to undercut his mission. For example, Hayes tried to discredit Silas Bent and T. B. Maury, who argued in lectures and popular articles that

FIGURE 9. Charles Hall. From C. Davis, *Narrative of the North Polar Expedition,*
frontispiece.

the Gulf Stream and another current, the Kuro-Siwa in the Pacific, fed the
polar sea with warm water, thereby keeping it open. By their calculations,
both currents entered the Arctic close to the surface of the ocean (the Kuro-
Siwa entered through the Bering Strait) and at their point of entry melted a
channel through the ice that surrounded the polar sea. Instead of slogging
through the torturous pack ice of Smith Sound, they claimed, explorers only
had to follow either one of the great currents to a "thermometric gateway"
that cleaved the great annulus of Arctic ice and offered smooth sailing to the
North Pole. The thermometric gateway theory threatened to make Hayes's
Smith Sound expedition obsolete.[25]

Hayes counterattacked in his own series of lectures and articles, direct-
ing his remarks to members of the scientific community. Bent and Maury
presented their theory in popular magazines such as *Putnam's* and *Galaxy*
and in lectures before historical and mercantile associations. Hayes aimed
his writings and lectures at Grinnell and other members of the AGS, the
audience that had proved so important to expeditions in the past. Because
Hayes's theory of the open polar sea did not differ much from that of Bent

and Maury, he relied on his superior authority as an explorer and man of science to persuade his audience. "The thermometric gateways to the Pole," he argued, "may take their place among those nicely and finely drawn fanciful phrases which, however available they may be in poetry, or in politics, are certainly quite misplaced in science; and with such phrases scientific men ought to have nothing to do, simply because they have no meaning whatsoever."[26]

CHARLES HALL

Hayes's greatest threat came from a rival explorer who had established himself outside the scientific community as a very different kind of Arctic authority. Like thousands of other Americans, Charles Hall had become interested in the Arctic just as Elisha Kane reached the zenith of his popularity as an explorer. As the thirty-eight-year-old publisher and editor of the *Daily Free Press* of Cincinnati, Hall had considerable exposure to the whirl of press stories about the Arctic in the 1850s. Hall's personal notebooks for 1857, the year of Kane's death, reveal his growing preoccupation with the subject. The readers of the *Daily Free Press* could not help but notice the new obsession of their editor as they read columns bulging with news and editorials about Arctic exploration. By 1859 Hall was suggesting to his readers that some of the Franklin party might still be alive in the Arctic. This seems astonishing. Fourteen years had passed since Franklin's departure from England. Rae's report of 1854 suggested that at least a third of Franklin's party had died in 1848, more than a decade earlier. Moreover, an expedition by the British explorer Francis Leopold McClintock had recently returned from King William's Land to report that most or all of Franklin's men had perished there. Such discouraging news did not faze Hall, who began planning his own expedition to King William's Land. "I felt convinced," he wrote later, "that survivors might yet be found."[27]

Echoing Dickens and other British boosters, Hall based his argument for pursuing the search for Franklin on character. Because he viewed the Franklin party as embodying the highest attributes of British character, it seemed unlikely to him that they all could have succumbed to the Arctic elements. "Is it possible," he said in one lecture, "that one hundred and five carefully selected officers and men, of one of the first nations of the world . . . should have perished?" In contrast to Dickens, however, Hall did not believe that character alone could explain the party's survival. He identified the Eskimos as their probable saviors. These "Iron Sons of the North," he suggested, had taken in the survivors of the Franklin party and taught them how to live in the Arctic. As he imagined it, the officers and crew had adapted themselves to the Eskimo way of life after they discovered that it would be impossible to escape from the Arctic. Hall sought support for this view among other explorers. Thomas Hickey, a member of the Kane expedition, testified that Hall's scenario was plausible. The men of the Kane expedition, Hickey reported, had faced a similar situation in 1855. If Kane had died before organizing an escape attempt, Hickey admitted, "I believe the men could never have found their way back to the Civilized World. . . . The Esquimaux life would have been finally adopted by us and I doubt not from the experience we had we might have lived to a good old age." Thus, Hall's theory emerged as a strange union of Rae's and Dickens's views, which had seemed so incommensurable a few years earlier. Continued hope for the Franklin party lay on twin pillars of character: that of the Franklin party as civilized men and of that the Eskimos as noble savages.[28]

Hall's dual defense of savage and civilized character underlay every element of his expedition. Anticipating that it might be difficult to find the same nomadic tribes that had taken in Franklin's men, Hall believed that extensive interviews with other local tribes would eventually lead him to the survivors. He reasoned that the same noble qualities of Eskimo character that had compelled them to rescue Franklin would ensure their candor in speaking about the missing men. In order to conduct such interviews, Hall planned on living with Eskimos near King William's Land to learn their language. He would thus prove by example that civilized men could adapt to the Arctic and its savage ways of life. On the surface, Hall's plan offered a compelling plot to American audiences who had been so drawn in by Kane and others: it would allow him to play the role of an Arctic detective, tracking down the Franklin survivors and liberating them from their long and savage polar captivity. At a deeper level, Hall's plan offered another compelling, if less clearly articulated, reason: it served as a experiment in which Hall, a man of the civilized world, would enter into a savage world in order to describe the borders and bridges that lay between the two.[29]

By making encounters with the Eskimos the centerpiece of his expedition, Hall capitalized on a growing interest in savage life among white Americans in the 1850s and 1860s. In particular, Hall's project struck a chord with Americans who had grown increasingly ambivalent about the decline of American Indians and the retreat of the western frontier. Earlier in the century, as white settlers struggled to gain control of native lands, portrayals of American Indians in popular and political texts tended to emphasize their cruelty and barbarism. Such accounts had served to justify the policies of Indian removal, which, along with warfare and disease, led to the rapid decline of Indian tribes east of the Mississippi in the 1830s and 1840s. Yet as the number of American Indians living in the East dwindled to a quarter of its former size, Euro-American representations of the Indians grew more benign and nostalgic, especially in the East. From Henry Wadsworth Longfellow's *The Song of Hiawatha* (1855) to Anna C. Miller's *The Iroquois; or, The Bright Side of Indian Character* (1855), white writers and artists closely identified Indians with the American wilderness, and their decline came to represent the retreat of nature in the face of white settlement. Within this context, even the Indians who successfully adapted to the encroachment of white settlement nevertheless relinquished their status as "the children of nature" because exposure to white civilization had contaminated their ways of life. Hall's plan to live with and report about so-called savages on the northern frontier, then, presented an excellent vehicle for urban white readers to enter into a raw experience of nature and primitive life at a time when it seemed doomed to extinction back home.[30]

Hall's theory of Franklin's rescue by Eskimos also played on a well-known popular literary genre: the captivity narrative. Since the seventeenth century, stories of whites captured by Indians had enthralled Euro-American readers. Publishers printed thousands of true-life and fictional captivity accounts in every region of the country. Although Hall's theory involved Eskimos, not Indians, it harmonized well with the central feature of captivity stories: white protagonists forced to endure and adapt to savage ways of life. His theory presented a more positive depiction of savages than did most captivity narratives, especially those written during the early nineteenth century, which emphasized the cruelty of Indians toward white hostages. But Hall's noble Eskimos fit in well with many captivity narratives of the mid-nineteenth century that gave a more balanced, even sympathetic account of savage life and mores. As with other depictions of American Indians, captivity narratives had changed to reflect the attitudes of many white Americans (especially in the East) who had mixed feelings about the settlement of the frontier. Perhaps such stories appealed to white readers because they offered a vicarious experience of savage life while still allowing the reader to identify with a civilized protagonist. Such appeals were consistent with Hall's mission, grounded as it was on optimism about the civilized as well as the savage character.[31]

Despite the congruity between Hall's plans and popular attitudes and literary genres, he had difficulty generating popular interest in his expedition.

However compelling Hall's theory about Franklin might have been, it did not convince audiences that a thirty-eight-year-old publisher with no Arctic experience could succeed in finding Franklin's men where so many others had failed. To demonstrate his commitment to and suitability for the mission, he began a program of physical training by camping out on a hill near the Cincinnati Observatory in order to acclimate himself to the hostile conditions of Arctic life. The only hostile elements Hall encountered were two drunken Irishmen who climbed the hill to badger him for whiskey. When Hall refused, they fired a shotgun at him. To the delight of local readers, the story of the would-be explorer's half-naked flight to the city ran in the next day's papers. Hall's preparations were garnering attention but not public support. He found more success in political circles, from Ohio governor Salmon Chase, Senator George Pugh, and others. These men encouraged Hall to move beyond his local, and rather comical, campaign and make a pilgrimage to the capital of American polar culture and patronage, the American Geographical Society in New York.[32]

A journey to New York in February 1860 brought Hall into direct competition with Isaac Hayes, who was promoting his own Arctic expedition. Hall at first failed to appreciate Hayes as a rival; after all, the goals of the two men were different: Hayes had no interest in finding the Franklin party, and his expedition to the open polar sea would take him hundreds of miles from Hall's field of search. Moreover, Hall had already struck up an amiable correspondence with Hayes and in the spirit of cooperation had shared the podium at the AGS with him to discuss their separate expeditions. But Hayes's experiences with the Kane expedition had shown him that explorers won and lost their campaigns at home, not merely in the field. Hayes correctly viewed the campaign at home as a zero-sum game: the two explorers competed for many of the same patrons and popular audiences even though they pursued different missions in the Arctic. As if to underscore the point, Hayes used their joint appearance at the AGS to meet Hall's expedition captain and induce him to leave Hall for his own expedition.[33]

Hall also found it difficult to compete with Hayes's reputation as a scientific explorer, which played well at the AGS and other established scientific societies such as the American Philosophical Society and the Academy of Arts and Sciences. Hayes's success in winning over Bache, Henry, and Agassiz had helped him gain money and support from associations up and down the East Coast. By contrast, Hall's lack of scientific training and his focus on Franklin dampened the enthusiasm of scientific audiences for his expedition. Where he found supporters, they were often unable to give him much financial aid, having already opened their wallets for Hayes. Grinnell served as a case in point. He had quickly become Hall's biggest and most visible advocate, but his support of the De Haven, Kane, and Hayes expeditions had left him unable to fund Hall's mission in kind. Having failed to raise the needed capital, Hall drastically revised his expedition. Instead of commanding a vessel, he now planned to secure passage on one: the Arctic whaler *George Henry*. Unable to pay wages for other expedition members, Hall was forced to travel without a crew, making his reliance on the Eskimos for all necessary information and support total. The fundraising campaign provided him with some provisions and a small whaleboat, with which he hoped to navigate the eight hundred miles (direct distance) to King William's Land. In May 1860, Charles Hall boarded the *George Henry* to begin his long-dreamed-of but much-altered expedition to the Arctic. Coverage of his departure by the *New-York Times* reveals the extent of his eclipse by Hayes. "On Tuesday last an Arctic expedition under the guidance of Mr. C. F. Hall set sail from New London, Conn. Intimations of this project have come to us from time to time, but the expedition of which Dr. Hayes is the head has so absorbed the attention of those interested in these matters that this announcement will doubtless take them by surprise."[34]

HALL'S EXPEDITIONS OF 1860 AND 1864

In the two-year expedition that followed, Hall failed to find Franklin's party or come anywhere near King William's Land. But he did succeed in validating the two premises of his expedition: the ability of civilized white men to adapt to Eskimo life in the Arctic and the capacity of the native Eskimos to provide honest and incisive information. As the *George Henry* cruised the southern coast of Baffin Land scouting for whales, Hall spent his time getting to know the Nugumiut Eskimos on the southeastern coast. A fierce gale destroyed Hall's small expedition boat, ruining any chance of making the long voyage to the King William's Land that year. Over the winter, Hall spent time learning the language and customs of the Nugumiut. With their assistance he made the only discoveries of his voyage. They informed him that Frobisher Strait, the long waterway that seemed to bisect Baffin Land, was actually a bay. This information, told so matter-of-factly by the Nugumiut, had long been a central mystery of Arctic geography in the West. Since the voyage of British explorer Sir Martin Frobisher in 1576, geographers had argued about the status of Frobisher Strait as the gateway to the Northwest Passage. Hall was also astonished to learn that, through oral tradition, the Nugumiut had maintained a detailed account of their encounters with Frobisher almost three hundred years earlier. Moreover, they provided Hall with the location and description of Frobisher's campsite, which he then visited to gather relics.[35]

In 1862 Hall carried home persuasive evidence to support his claim that Eskimos' character and knowledge made them essential in determining the fate of the Franklin expedition. As Hayes had used evidence of the open polar sea from Kane's expedition to establish a rationale for a new expedition, Hall used the Frobisher relics to lay the foundation for his second expedition to find the Franklin party. They demonstrated the reliability of the Eskimos as witnesses of events in the Arctic, directly refuting the representations put forward by Kane and Dickens. In 1863, as he prepared his narrative for publication, he made this link explicit in his preface: "The reader will perhaps wonder why so much importance was given to the discovery of the Frobisher relics . . . the discovery of these remains, and the tracing of their history among the Esquimaux, confirmed, in a remarkable manner, my belief that these people retain among them, with great positiveness, the memory of important and strange incidents; and as their traditions of Frobisher, when I was able to get at them, were so clear, I am persuaded that among them may be sought . . . the sad history of Sir John Franklin's men."[36]

To strengthen his case with his audiences, Hall also returned with a Nugumiut family with whom he hoped to persuade Americans of the "great positiveness" of the Eskimo race. Hall chose this family in part because of their ability to conform to his new image of Eskimos. Tookoolito, her husband Ebierbing, and their infant son Tukerlikto were in no way typical members of the Nugumiut tribe. They already had considerable experience with European culture, having traveled to England at the urging of a British captain. They spent almost two years there, hobnobbed with the British aristocracy, met Queen Victoria and Prince Albert, and learned British etiquette and tastes. Their fluency in English and familiarity with Western customs made them perfect for the task that Hall hoped to accomplish: reinforcing the image of Eskimos as a trustworthy people, and in the process, strengthening the argument for a second expedition.[37]

Hall recognized that presenting Tookoolito and Ebierbing as too civilized threatened to reduce their appeal with popular audiences. Kane, Dickens, and others had cultivated an image of the Eskimos as a primitive people. For white Americans who lamented the disappearance of Indians from the frontier, Eskimos represented a last bastion of savagery as yet untouched by the forces of civilization. Thus, as he brought the Eskimos on stage after his lecture at the AGS, Hall was careful not to undermine their savage appeal, balancing his positive portrayal of them with displays designed to highlight their authenticity as Arctic primitives. For example, even though Tookoolito and Ebierbing were comfortable wearing Western clothes (Hall had first encountered Tookoolito on Baffin Land wearing crinoline and a bonnet), he had Tookoolito sew sealskin suits for their appearance at the AGS. As they walked onto the stage after Hall's lecture, they carried other implements of Arctic life that confirmed their status as primitives: bow and arrows, fish spears, and two barking Eskimo dogs in full harness. Only after this dramatic introduction did Tookoolito and Ebierbing demonstrate their less "savage" side, bantering with surprised audiences in British English.[38]

Hall used the strategy with great success in other venues as well, bringing the Eskimo family on stage with him in more than a dozen lectures in New England and New York. Hall's interest in promoting the high character of the Eskimos did not prevent him from exploiting Tookoolito, Ebierbing, and Tukerlikto as human curiosities. Capitalizing on the Eskimos' popularity among eastern audiences, Hall arranged with Boston and New York sideshows to have them appear on their own. Their popularity at Barnum's Museum in New York led Hall to extend their contract; there they received top billing over other attractions such as Old Adams' California Dancing Bears and The Albino Family. But such tours, in addition to the many lectures, increased the Eskimos' exposure to diseases for which they had little or no resistance. By the winter of 1862–1863 all three suffered from illnesses, and Tukerlikto died the following spring.[39]

The success of Hall's lectures about the Eskimos, however, could not overcome the difficulties that the Civil War posed for Arctic expeditions. The war impeded Hall's efforts just as it had for Hayes. Although Hall's wartime lectures had proved to be far more popular than his rival's, they did not generate sufficient revenue to set up a second expedition. Unable to raise money among the polar elite in New York, Hall shelved his plans for commanding a ship bound for the shores of King William's Land. As he had with the first expedition, Hall cobbled together a modest mission with the help of a few Arctic stalwarts such as Grinnell. He again entered the Arctic as a lone passenger aboard an Arctic whaler, in this case, the *Monticello*. The whaler would head for Repulse Bay, on the northern shores of Hudson Bay. There he would attempt to enlist the Repulse Bay Eskimos in the long overland journey to King William's Land to find the (now graying) Franklin survivors.[40]

Hall's second expedition lasted five years, during which time he made a number of overland journeys with the cooperation and support of the Repulse Bay Eskimos. In 1869 he succeeded in reaching King William's Land. Together, the testimony of local Eskimos and the bleached and scattered bones on the island finally convinced him that all of Franklin's men had perished in the Arctic. Despite this disappointing conclusion, however, Hall dramatically confirmed his original claims: that Eskimos possessed knowledge and skills that were invaluable to extended Arctic exploration and that whites could successfully explore the Arctic by adopting Eskimo ways of life. Yet his success in resolving the fate of Franklin's party also deprived Hall of his principal rationale as an explorer. For almost ten years the Franklin party had been the object that had shaped not only Hall's mission but his persona and popular campaign as well. As he returned home, Hall fixed his sights on the only other compelling Arctic targets: the polar sea and the North Pole.[41]

HALL'S EXPEDITION TO THE NORTH POLE

The expense of an expedition to the North Pole forced Hall to change his campaign in order to appeal to new audiences. On his previous expeditions, when he failed to raise enough money to buy a ship and hire a crew, he had used Arctic whalers as transport. But whalers did not ply the dangerous waters of Smith Sound or the polar sea. A credible expedition to the North Pole required considerably more overhead: he needed to purchase and heavily reinforce a ship, hire a crew, and buy two or three years' worth of equipment and provisions. His previous expeditions had been organized on shoestring budgets, funded in large part by Grinnell and a handful of other patrons. But losses incurred during the Civil War made it impossible for Grinnell to back the expedition himself. And popular lectures, by themselves, could not cover the costs of a large expedition. Instead, Hall hoped to convince Congress that it should become the major patron for his new expedition. The Civil War had undermined interest in Arctic exploration, but it had increased the power of the federal government. Twenty years earlier, Congress had taken enough interest in the Franklin search to place the Grinnell expedition under naval command. A few years later, it had supported a rescue expedition to search for Kane. With the war behind them, Hall hoped to revive federal interest in the Arctic. At small-town lyceums, it had been enough for Hall to thrill his audiences with Tookoolito, Ebierbing, and their retinue of dogs. But in order to charm this new patron, Hall needed to tell different kinds of stories.

Washington elites pressed Hall to define the national value of his expedition as well as his own abilities as an explorer.[42]

In the 1850s and 1860s, the Washington scientific community, in particular the Smithsonian, the Coast Survey, and the Naval Observatory, increasingly served as the gatekeepers for federal support of expeditions. Hall's greatest challenge lay in winning over these elites. Hall set to work trying to convince Henry, Bache, and other luminaries of the value of his expedition. It proved to be a difficult sell. Ever since his first expedition in 1860, these elites had been skeptical of Hall's methods and goals as an explorer. Because Hall had focused on rescuing the Franklin party rather than on scientific or geographical discovery, they had viewed his voyages less favorably than Hayes's expedition to the open polar sea. Moreover, Hayes had also cut a superior figure as a scientific explorer. Whereas Hall had no scientific training at all, Hayes could boast of being a university-trained physician. Apart from his lectures before the AGS, which had been engineered in part by Grinnell, Hall had not been welcomed at prestigious scientific societies as had Kane and Hayes.

As he attempted to foster a relationship with scientific societies, Hall found himself rebuffed. For example, when he approached the Smithsonian to offer a lecture about his expedition, as had Kane, the Smithsonian turned him down, citing a "general rule against pay lectures." Similarly, these agencies proved reluctant to lend Hall equipment for his expeditions, although they had lent equipment to Kane and Hayes enthusiastically. Bache's staff at the Coast Survey informed Hall that equipment he requested was broken. Bache later agreed to help Hall, but only "as an individual" and not in Bache's capacity as superintendent of the Coast Survey. Henry proved more forthcoming than others in his reasons for denying Hall equipment: "Scarcely any results could be obtained," he wrote him, "unless some one properly educated for the business of observation should devote his whole time to the instruments." Some critics made their charges public. One of the men considered for the position of expedition naturalist, Dr. David Walker, claimed that Hall's new interest in science was merely a ruse to gain federal support for his expedition. In an article published in the *Overland Monthly*, Walker quoted Hall as telling him: "I do not care one cent for science; my object is to place my foot upon the Pole."[43]

Lacking credibility as a man of science, Hall found it difficult to use his considerable experience in the Arctic to gain purchase with scientific audiences. By contemporary standards, Hall's narrative, *Life among the Esquimaux*, was an impressive ethnographic account. First published in

London in 1864, the narrative offered a detailed, if judgmental, portrait of the Frobisher Bay Eskimos. Perhaps in order to increase its appeal to scientific audiences, Hall changed the title of the American edition to *Arctic Researches and Life among the Esquimaux*. Yet judging by the absence of discussion about the work at scientific meetings and the lack of reviews in leading professional journals, this work went unnoticed in American scientific circles. This lack of attention can be explained, in part, by the secondary status of ethnography among these elites. In the 1860s, Washington men such as Henry, Bache, and Maury remained more concerned with physical and terrestrial sciences than with ethnography. It appears that Hall's low credibility as a scientific explorer also discouraged men of science from taking the work seriously as an ethnographic text. Although Hall's willingness to cast off civilized behaviors in order to live with Arctic "savages" had thrilled popular audiences, it also reinforced images of him as an uncultivated man, increasing the social distance between him and the genteel community of men from whom he now sought support. Hall's critics were quick to seize on his success in living with the Eskimos as evidence of his unsuitability for commanding a scientific expedition to the North Pole. "Captain Hall's only acquirements, so far as I know," David Walker wrote, "are the possession of an indomitable energy, a boundless enthusiasm, and a capacity for uncivilization." The *Atlantic Monthly* praised *Arctic Researches* but at times found Hall's appreciation for Eskimo life "naïve" rather than scientific. The comment echoed the language of Kane and others who had described the Eskimos metaphorically as children. Now Hall found himself appraised with similar language.[44] Lecturing to the APS on the value of the search for the North Pole in 1868, Thomas Kane, the explorer's brother, stressed the importance of science to polar expeditions. Although Kane's words did not refer to Hall specifically, they applied directly to the predicament Hall faced the following year: "What good will it do for an ignorant Indo European to scamper up and scamper down—to stand on the polar hub or axis even, and relate that he was there? Little more than to have a savage do the same thing—a little more than a dog ... the truth is, man in the nineteenth century, without philosophical instruments, without educated powers of observation, is less than a man."[45] Hall's failure to attract scientists' support for his expedition presented a golden opportunity for Hayes. To Hall's shock and dismay, Hayes traveled to Washington in 1870 to present himself to Congress as an alternative candidate for the North Pole expedition. Unable to compete with Hayes within the scientific community, Hall turned to popular and political audiences directly, hoping to convince them that his

Arctic experience made him a better qualified expedition leader than Hayes, who had not commanded an expedition in ten years. Eager to demonstrate that his previous expeditions had given him more than a "capacity for un-civilization," Hall now spoke about them using the language of science and professionalism. Representing Arctic exploration as a highly specialized activity, he portrayed his earlier expeditions as constituting a rigorous course of instruction. "Having now completed my Arctic collegiate education," he wrote to the president of a teacher's institute, "I hope to spend my life in extending our knowledge of the earth up to the spot which is directly under Polaris." The popular press picked up this metaphor and used it in its coverage of Hall's lectures. Reporting on a speech he gave in Brooklyn, the *New York World* wrote: "He has gone through a full course in the Arctic college." Hall's testimony before Congress also framed his experience as consistent with the goals of scientific exploration. He told the Committee on Foreign Relations that there was considerable precedent for scientific expeditions carried out under the command of a man of experience, rather than a man of science. "I am not a *scientific* man," Hall testified. "Discoverers seldom have been. Arctic discoverers—all except Dr. Hayes—have not been scientific men. Neither Sir John Franklin nor Sir Edward Parry were of this class and yet they loved science and did much to enlarge her fruitful fields." In fact, Hall's supporters suggested that such a lack of training would make Hall a better and less prejudiced observer of Arctic nature. "Hall was all the more fit for his undertaking," one Hall loyalist later wrote, "because he was not wedded to any particular branch of science, but that like Livingstone, he devoted all his energies to geographical discovery." Hall's attempts to frame his earlier expeditions as a scientific education played off the broad public support for "self-culture." More specifically, it resonated with the claims that Kane's eulogists had made in praising him for his self-education rather than his formal degrees.[46]

Hall also used social connections outside of the scientific community to gain access to members of Congress. His two earlier campaigns had brought him into contact with influential men in Cincinnati and New York. He now used this network of powerful friends to secure interviews with two Ohio legislators who agreed to promote his expedition in Congress. Later he asked his friends to give him letters of introduction to other members of Congress. Hall even managed to meet with President Grant to discuss the expedition, bringing Tookoolito and Ebierbing along with him for the occasion. In contrast to Hall's earlier rhetoric, which had dwelt on the character of the Eskimos, his pitch for his new expedition used the language of nationalism. "The day has come," he wrote one supporter, "when the great problem of ages on ages must be solved under the stars and stripes." In a lecture at Lincoln Hall attended by President Grant and Vice-President Colfax, Hall reiterated this theme, and his calls for federal support brought roars of applause.[47]

Hall's success in framing his Arctic experience as specialized training rather than an exercise in "going native" won Congress over. By appealing directly to popular and political audiences, he made an end run around Hayes and the elite scientific community. In 1870 Congress appropriated $50,000 for his expedition to the North Pole. Yet he did not escape scientific oversight of his journey. Congress directed Joseph Henry and the National Academy of Science to advise Hall about scientific plans and personnel. Henry assisted him willingly and without complaint. In his final report to Hall (and to Congress), Henry stated that the NAS fully approved of Hall and his mission. Yet one detects a whiff of disapproval in the report. "It is evident," Henry concluded, "that the Expedition, except in its relation to geographical discovery, is not of a scientific character." Moreover, he seemed anxious to distance himself from the potential failure of the scientific mission. "Should the results not be commensurate with the anticipations in regard to them," he wrote, "the fact cannot be attributed to a want of interest in the enterprise or to inadequacy of the means which have been afforded." Walker, writing after Hall set sail, could speak more candidly: "Much, if not all, of this lukewarmness was consequent upon the selection of the leader, and the resultant sacrificing of the positive advantage of scientific research to a vain-glorious attempt to flaunt the Stars and Stripes."[48]

Although Hall's expedition set a new record for "farthest north," in all other respects it proved disastrous. By the time his ship, *Polaris*, reached Greenland, he was already embroiled in quarrels with the scientific team appointed by Henry and the NAS. The two members of this team, Emil *German* Bessels and Frederick Meyer, refused to follow orders that they believed interfered with their duties. Meanwhile, unusually good ice conditions allowed *Polaris* to sail almost to the top of Smith Sound, at latitude 82° north, before being pushed back by the encroaching pack ice. After returning from a short trek up the Greenland coast, Hall drank a cup of coffee and immediately suffered a violent seizure. The suddenness of the attack led him to believe that he might have poisoned by one of the crew. For a week his condition deteriorated, and he became delirious. After a short recovery, Hall lapsed into a coma and died on 8 November 1871. His death further eroded discipline aboard the ship. The following year *Polaris*, already damaged by

Was Hall poisoned?

her winter harbor, drifted into Smith Sound with the pack ice. When a fierce storm threatened to crush the ship in October, the crew started unloading stores and equipment on the ice. In the fury of the gale, the ship broke away from the floe, stranding nineteen of the crew members there. The fourteen remaining crew members grounded *Polaris* on the Greenland coast. With the help of the Etah Eskimos (who had now grown accustomed to rescuing American explorers) the *Polaris* party survived the winter, making their way down the Greenland coast. They were picked up by a Scottish whaler the following summer. Meanwhile, the nineteen stranded members of the expedition (including Tookoolito and Ebierbing) gradually floated south. This party, too, would have perished without the skill of the Greenlander Hendrik (who had joined Hall's party as he had Kane's and Hayes's) and Ebierbing, who continually secured game during the long winter. At the end of April the following year, an American sealer picked up the crew off of the coast of Labrador, after they had spent more than six months on the ice.

But the ordeal was not over. The navy convened a Board of Inquiry to investigate the suspicious circumstances of Hall's death. The board ruled that Hall died from natural causes (a finding that has since been called into question), but it also laid bare the complete breakdown of authority that followed the commander's demise. Crew members and officers drank heavily, following the lead of the ship's captain, who frequently appeared drunk on deck. The Arctic had not brought out the crew's best character. Hall had hoped to gain honor for himself and the United States, but his expedition seemed to have had the opposite effect. As one member of the expedition wrote, the affair had left him "utterly disgusted...with myself, the country, everything in it, and everybody that has a fancied interest in it."[49]

CONCLUSION

Hayes's and Hall's expeditions were plagued by mishaps and misfortunes, capped off by the death of Hall and the disintegration of his expedition in 1871. But these fiascos obscure the success each explorer achieved in appealing to specific American audiences. For example, one exploration historian has expressed surprise at Hayes's success with the scientific community given his limited experience in the Arctic. "It is quite astonishing that he was able to impress so many geographical, philosophical, and scientific societies, and attract the support of so many distinguished and influential men (including Henry Grinnell), for the sum total of his Arctic 'exploring' experience had been a twenty-five mile trip to the edge of the Inland Ice, and a ten-day journey to the west coast of Kane Basin." But Hayes's success is only surprising if we assume that men of science evaluated him on the basis of his field experience. For Hayes such acts of evaluation took place at home, in the lectures halls and living rooms of the scientific elite. Here he succeeded in gaining support not so much by enumerating his feats in the Arctic as by demonstrating his fluency in and fidelity to issues that were important to men of science. Hall found success with popular audiences for much the same reason, connecting his expedition to issues that were of proven interest to them. By connecting his Franklin mission to white middle-class interest in savagery and captivity stories, he overcame the handicap of lack of experience. Later, however, as he competed with Hayes to lead an expedition to the North Pole, Hall was able to frame his now-considerable Arctic experience as a form of scientific "self-culture" that played well with popular and political audiences. Yet for all of their successes, Hayes and Hall were unable to unite different groups in support of their expeditions. This failure had as much to do with events beyond their control—Rae's report and the onset of the Civil War—as it did with their actions in the Arctic. It should be remembered that Kane had returned in 1855 from a voyage to Smith Sound without specimens, data, or a ship and received a hero's welcome. Six years later, however, Hayes returned from a similar mission to a war-weary public who viewed his mission as a wasteful antebellum extravagance.[50]

The difficulty of rallying different American groups in a common Arctic cause also reflected changes in the groups themselves. The 1860s and 1870s witnessed a dramatic shift in the relations between scientific elites, amateurs, and the general public. Demand for scientifically trained personnel in industry, agriculture, and education aided elites in promoting scientists as a professional class. The inclusive scientific societies that had so warmly received Kane in the 1850s had begun to establish new guidelines for its members and officers. If the untutored Hall felt snubbed by such elites, he was not alone. Amateur members of the AAAS and other scientific societies were also grumbling as ruling committees sought to exclude them from positions of power. In short, the fault line separating serious and popular approaches to Arctic exploration may have become more visible after the appearance of Rae's report, but its causes lay deeper in the tectonic movements of science as a cultural practice.

The unique demands of their target audiences led both men to represent themselves and their missions in very different ways: Hayes portrayed himself as an explorer of cultivation and scientific sensibility, cut from the same cloth as Elisha Kane. By contrast, Hall portrayed himself as a man of the frontier whose zeal to find Franklin had compelled him to live like a savage. Such different self-representations, in turn, helped delineate different visions of the Arctic and its peoples. Hayes emphasized the Arctic as a place of sublime processes that touched the rational and scientific as well as the subjective and emotional faculties of the explorer. Establishing the interaction between himself and the Arctic as the central relationship of his writings, Hayes played down the importance of the Eskimos to his expedition in the same manner as Kane had. More often than not, the view of Eskimos as "a negative people" offered Hayes a way of contrasting his own high character and that of his crew. By contrast, Hall made his interaction with the Eskimos the central feature of his narratives as well as the foundation of his expedition plans. Consequently, Hall's reveries about the Arctic often took a back seat to his descriptions of individual Eskimos and their ways of life. Not surprisingly, they emerge as more active and more sympathetic agents than those in Hayes's accounts.

Hall's rebellion was a quiet one. He never abandoned the mantle of civilized character. Nor did his appreciation of Eskimo life ever challenge the supremacy of American values. But his campaigns represent an important departure from Kane's. Hall's success demonstrated the receptiveness of his audiences to a new kind of rhetoric, one that reveled in, rather than recoiled from, the primitive frontier. He largely ignored science. And he gave voice to themes that foreshadowed a dramatic shift in exploration rhetoric that would emerge in the years after the Civil War, a time when explorers would more forcefully challenge traditional ideas about manliness, science, and progress. Hayes's and Hall's careers thus reveal new dilemmas facing Arctic explorers as public figures. How did one best promote polar voyaging? As missions of science or of manly character? Or as something entirely different, a flight from civilization? Answering these questions would shape the course of polar campaigns and expeditions for the rest of the century.

———— ✳ ————

CHAPTER FOUR

Dying Like Men

Adolphus Greely

IN JUNE 1884, Commander Winfield Scott Schley cruised the waters of Smith Sound searching for Adolphus Greely and his missing party of American explorers. Greely had been in the Arctic for three years, establishing a scientific station at Lady Franklin Bay as a part of the International Polar Year. Two attempts to relieve Greely, in 1882 and 1883, had failed, and Schley's expedition represented the last reasonable chance of finding the Greely party alive. When one of Schley's men discovered a note from Greely giving his location at Cape Sabine, Schley sent John Colwell and a small party to find him. Arriving at the site, Colwell found Greely along with six other emaciated men, survivors of the original party of twenty-five. In his narrative of the rescue, Schley described the scene: "Colwell crawled in [the tent] and took him by the hand, saying to him, 'Greely, is this you?' 'Yes,' said Greely in a faint, broken voice, hesitating and shuffling with his words, 'Yes—seven of us left—here we are—dying—like men. Did what I came to do—beat the best record.' Then he fell back exhausted."[1]

Schley was not on the beach and relied on the reports of his men to piece together the events. Yet his narrative, published almost a year after the return of the survivors, soon gained authority as a true-life account of the dramatic rescue. The *New York Herald* excerpted it liberally in its reports about the expedition. Later reminiscences, such as *Munsey's Magazine's* 1895 account of the expedition, presented the scene exactly as it had appeared in Schley's narrative. Even David L. Brainard, one of the seven survivors of the Greely party, used Schley's dialogue word for word in his 1929 account of the expedition, as well as in a more candid narrative that he published in 1940.

Yet others on the beach recalled the meeting differently. One member of the rescue party reported that Greely first asked him "if we were Englishmen." Another remembers Greely chastising them. "If we've got to starve ... we can starve without your help ... we were dying peacefully until you came." Maurice Connell, one of Greely's men, was unconscious at the time of the rescue. Yet for him the published account of Greely's words did not ring true. In the margins of Schley's book he wrote: "'Give us something to eat!' more probably."[2]

Whether Greely, delirious and close to death, uttered his pithy remarks is unclear. What is clear is that the scene Schley described offered a far more respectable image of Greely and his party than the ones that circulated for months in the popular press on Greely's return. Immediately after the return of Schley's rescue expedition, the *New York Times*, the *Chicago Tribune*, and other papers deluged readers with lurid stories about the Greely party's demise on Cape Sabine. They uncovered Greely's execution of a man for stealing food. They reported on rumors of cannibalism among the party and discussed charges by Greely's men that he was inept as a commander. Schley's account did not erase the impact of these revelations, but, arriving in the wake of the reports, it offered a means of capping the well of controversy, as its extensive use suggests.

The controversy that engulfed Greely after his return eclipsed his expedition's scientific work. Billed as the most ambitious research mission ever sent into the Arctic, it marked instead the end of serious collaboration between scientists and Arctic explorers in the nineteenth century. In the decades to follow, explorers occasionally promoted their voyages as research expeditions, but their words had little bearing on their expeditions or their campaigns. New patrons of Arctic exploration freed explorers from having to appeal to the scientific community to raise funds or lobby Congress. From the point of view of scientists, Greely and other explorers had abandoned their research missions in order to give allegiance to new masters, private patrons and press moguls who cared little about Arctic science. Schley's account of the rescue only underscored the point. Greely declared to Colwell that he "did what [he] came to do—beat the best record," but he had not entered the Arctic to do anything of the kind. In fact, organizers of the Greely expedition had hoped quite the opposite: that Greely and his men would turn their backs on dangerous and irrelevant dashes to the North Pole, focus on methodical research, and embrace the collaborative spirit of the International Polar Year. To his credit, Greely carried out much of this research at first, but he eventually turned his attention to the geographical dashes so disapproved of by international organizers. Greely's words at Cape Sabine, true or apocryphal, only confirmed suspicions that science had also been a casualty of his expedition.

The claim that the Greely expedition marked the end of explorers' serious collaboration with scientists stands at odds with most historical accounts. Because Greely brought back the most comprehensive and systematic set of observations ever produced by Americans in the Arctic, historians have often held it up as a sign that U.S. scientific exploration had come of age. For the historian of science A. Hunter Dupree, the Greely expedition "laid the groundwork for a really scientific interest in Arctic and Antarctic problems." The wealth of data collected by Greely and his men has led William Barr to revisit the events of the expedition in hopes of illustrating its scientific importance. The expedition historian A. L. Todd agrees, calling Greely's official narrative "one of the most important source books of Arctic data available to the world of science." Focused on quality of the data, however, these works leave unexamined the reactions of scientists, the press, and the public back home. Fixing our attention on these, we see a different picture emerge: a diminished role for science in Arctic exploration, waning collaboration between explorers and researchers, and a decline of scientific rhetoric in expeditionary campaigns.[3]

The reasons for this change extend beyond the expedition. Greely and his rescuers may have sown the seeds of controversy by their actions in the Arctic, but these actions only bloomed into scandal because of important cultural and institutional changes back home. That science fell victim to scandal after Greely's return reflected a new trend in newspaper journalism that put a premium on critical, often sensational, reporting. This was a far cry from the 1850s, when reporters generally avoided controversy in their attempts to portray Arctic explorers as American heroes. By the 1880s, however, writers proved far more willing, even eager, to expose expeditions' scandals at explorers' expense.

The estrangement of explorers and scientists also grew out of changes in the patronage of Arctic exploration. Whereas Henry Grinnell had encouraged his explorers to campaign as a way of raising funds and creating broad coalitions among scientists and the public, the deep pockets of new patrons such as the *New York Herald* made such actions unnecessary. Flush with funds and backed by the promotional power of their patrons, explorers had little need to campaign. Greely took his orders from the Army Signal Corps, not the popular press, but many of the effects were the same. With his expedition already organized and underwritten by the corps, Greely felt no incentive to write, lecture, or rub elbows with scientists or other groups in the months before his departure. As a result, he established few of the personal bonds with scientists and others that had so benefited Elisha Kane and helped insulate him from his critics.

THE PRECEDENT OF SCANDAL: THE *JEANNETTE* EXPEDITION

The threats posed by changes in patronage and press coverage had become visible before the demise of the Greely expedition. In the winter of 1881, as Greely and his men settled in for their first winter in their fort on Lady Franklin Bay, Americans back home learned about the demise of the *Jeannette* expedition in the Arctic. In many ways, the scandal that engulfed it set the stage for Greely's hostile reception by the press four years later. Sponsored by *New York Herald*'s publisher, James Gordon Bennett, the *Jeannette* expedition signaled a new and commercially lucrative relationship between explorers and the popular press. In the 1850s and 1860s, publishers had exploited the popular appeal of Arctic exploration but left the planning and promotion of expeditions to others. Bennett, by contrast, took these tasks for himself, making the *New York Herald* both the publisher and the promoter of expeditions to the Arctic and elsewhere. Bennett did not leap into exploration planning unprepared. His Arctic enterprise fit into a broader business philosophy: making the news was the best way to cover it.[4]

Bennett's *Herald* broke new ground, packaging its expedition coverage with sensational writing, bold headlines, and eye-catching illustrations, all of which helped catapult it into the lead in its fight for circulation in New York. Shortly after inheriting the *Herald* from his father, Bennett started sending his reporters on far-flung expeditions as a way to gain first-hand, exclusive accounts of expedition life. The most celebrated of these took place in East Africa, where, in 1871, *Herald* reporter Henry Stanley found the long-missing British explorer David Livingstone, a feat that brought Stanley and the *Herald* to world attention. *Herald* reporters also voyaged with relief vessels searching for Charles Hall in 1873. They ventured into the Northwest Passage with the *Pandora* expedition in 1875 and traveled to King William's Land with Frederick Schwatka to gather relics of the Franklin party in 1878.[5]

The success of such ventures prompted Bennett to sponsor his own expeditions. After briefly considering sending Henry Stanley to the North Pole, Bennett settled on George Washington De Long, a naval officer who had found a passion for exploration in the search for Hall's expedition to the pole. Encouraged by new reports of a "thermometric gateway" to the open polar sea through the Bering Strait, Bennett set to work planning an expedition and promoting it in the *Herald*.[6] Like earlier campaigns by Kane and Hayes, the open polar sea was both the object of study and the means to an end: in this case, reaching the North Pole. But Bennett's campaign differed from earlier ones in an another important, albeit subtle, way: it bolstered the reputation of the expedition rather than the character of its commander. Because Bennett assumed all of the expenses, De Long dispensed with the public lecture tours that had helped establish Kane, Hayes, and Hall as well-known figures. The *Herald*'s power in shaping the news, and Bennett's record of pro-navy editorials, had already given the publisher considerable influence with members of Congress and the navy. Thus, when De Long sought to put his expedition under naval command, he looked no further than Bennett in order to pressure Congress into nationalizing the expedition. Whereas Kane and his scientific allies had spent months lobbying Congress— ultimately unsuccessfully—to place their expedition under naval command, Bennett secured congressional action for De Long in a few weeks.[7]

Bennett's success in putting together the *Jeannette* expedition so quickly, without the support or advice of the scientific community, had serious consequences. In the short term, it gave him a free hand to set the agenda for the expedition and promote it within the pages of the *Herald*. In the long term, it hindered the scientific work of the voyage and left the expedition without a base of support in the scientific community. The low priority afforded science became clear soon after the *Jeannette* sailed through the Bering Strait. Instead of finding an open channel to the North Pole, the *Jeannette* found itself locked fast in pack ice. This grim discovery would prove to be the scientific highlight of the cruise. Never to sail again in open water, the expedition put to rest the theory of a thermometric gateway to the North Pole. Other scientific projects came to nothing. Raymond Newcomb, a young taxidermist aboard *Jeannette* who served as its science officer, dutifully shot and stuffed any game that wandered too close to the ship. But Newcomb's social isolation, fueled in part by the party's lack of interest in science, hampered him in his duties. Inexperience hindered other aspects of scientific work during the cruise. The slow drift of the *Jeannette* offered an excellent opportunity for sustained geomagnetic observations. But after an astronomical station on the ice was destroyed in an upheaval of the ice pack, De Long canceled construction of another one for fear of losing valuable instruments. Even basic navigational observations became difficult after sextants and chronometers were affected by the cold. After falling out with Jerome

Collins, who doubled as the *Herald*'s correspondent and the ship's physicist, De Long relieved him of his scientific duties, ending any hope of substantial scientific findings to bring home.[8]

These scientific failures were overshadowed by a much larger catastrophe. Almost two years after the *Jeannette*'s departure from San Francisco, pack ice stove in the ship's hull four hundred miles from the shores of Siberia. As the *Jeannette* slipped under the ice, De Long and his men marched slowly south, reaching open water three months later. The party sailed for land in three whaleboats until a fierce gale separated them. One boat was lost. The other two made landfall on the vast Lena delta. De Long's boat came ashore a great distance away from the second boat, commanded by *Jeannette*'s chief engineer, George Melville. Melville quickly found aid among local natives of the delta, but the men of De Long's party trudged south without assistance, accurate maps, or adequate provisions. De Long and all but two of his men died from hunger and exposure. Unaware of De Long's plight, Melville waited for his men to recuperate before starting a search.[9]

Melville's decision to delay a search sparked a storm of controversy back home. As news trickled out of Siberia about the death of De Long and nineteen of his men, criticism of Melville in the press was prompt and severe. Soon reporters leveled criticism more broadly at expeditions to the North Pole. The *New York Times* called the search for the North Pole "a criminal waste of money and a reckless risk of life." The *Nation* lamented the victims of the *Jeannette* expedition, whose "lives were uselessly sacrificed." Having watched the *Herald*'s extraordinary success in capitalizing on Henry Stanley's expedition in search of Livingstone, as well as its reporters' forays into the Arctic, competing newspapers were already inclined to view the *Jeannette* expedition as a publicity stunt. Thus they were less willing to give Melville and the survivors the same benefit of the doubt that they had given Kane in the 1850s.[10]

Bennett's reaction to the tragedy only confirmed his commercial priorities regarding the expedition. After receiving word from Melville by telegraph, Bennett immediately sent reporters to Siberia in order to capitalize on public interest in the tragedy. Back home, his newspaper published reports and editorials that, though more subdued in tone, also questioned Melville's actions in leaving the Lena delta. Reports from Siberia further stoked the flames of controversy. The *Herald* published interviews with John Danenhower, one of the survivors of the *Jeannette*, that questioned the command abilities of De Long and Melville. Later, *Herald* reporter John Jackson disinterred the remains of De Long and his boat crew in the Lena delta. The paper published

Jackson's report along with illustrations of the grave site. Although Bennett attempted to suppress the most damaging revelations of the expedition, he made the most of the controversy when it presented itself.[11]

Tragedy made the *Jeannette* expedition a popular sensation at a time when interest in Arctic exploration appeared to be in decline. Public interest in the region after the Civil War had never matched the initial interest generated by Kane's expedition of 1853. In fact, only a few months before news of the *Jeannette* story broke, the *New York Times* had observed, "Public opinion has been so lukewarm as to the fate of the many brave men on the *Jeannette*." Few publishers failed to notice how tragedy and scandal breathed new life into the story in 1882. At the same time, the rise of investigative and sensational genres of reporting in the 1870s gave cover to publishers and editors who wanted to focus on tragedy and scandal instead of manly heroism.[12]

As for the role of science in Arctic exploration, the *Jeannette* expedition generated different reactions. For some, it raised the bar for new voyages by demanding scientific results that would justify putting American lives in danger. The expedition's thin veneer of research did not merit such risks. "The *quantum* which is gained for science," Senator George Edmunds of Vermont declared, "is entirely outweighed by the suffering and loss of life and property." William Healy Dall, who frequently commented on exploration for the *Nation*, took away a similar lesson from the tragedy. "[P]opular weariness is justified in the case of many Arctic expeditions with which we are familiar. The only sound ground upon which the public may properly be taxed for such voyages is that of the advancement of scientific research." In Dall's view, only Arctic science could renew the public's trust.[13]

Publishers took away a different message. The volley of successful news stories that followed in the wake of the tragedy indicated that scandal offered a better remedy for "popular weariness" than did science. In their own promotion of the *Jeannette* expedition, book publishers appeared to come to the same conclusion. Of the four narratives written by various members of the expedition, only one made any attempt to examine its scientific findings. Dall wrote disparagingly of an account purportedly edited by ship's naturalist Newcomb, noting that "a careful examination has failed to discover any contribution of importance either to the history of the expedition or the natural history of the polar regions." Yet despite their critiques of Arctic exploration, the *Herald*'s competitors soon entered the expedition business themselves, featuring a variety of reporter-explorers who now made little pretense to scientific exploring in their journeys or their published reports.[14]

THE GREELY EXPEDITION

In its rigorous focus on science, Greely's mission to Lady Franklin Bay appeared to be an antidote to the *Jeannette* expedition. Organized by the Army Signal Corps, the expedition planned to conduct research in a more methodical and comprehensive manner than any previous American voyage. The plan involved sending a party of army personnel to Lady Franklin Bay, on the northeastern shores of Grant Land (Ellesmere Island). In conjunction with other polar stations set up by other countries as a part of the International Polar Year, the American party would measure atmospheric pressure, temperature, relative humidity, wind speed, and magnetic variation in hopes of establishing comprehensive meteorological and geomagnetic models of the Arctic, and by extension, the globe.[15]

The plan for the multinational research program originated with the Austrian explorer Karl Weyprecht. After returning from his own Arctic expedition in 1874, Weyprecht began arguing that the emphasis that explorers placed on geographical discovery, first of the Northwest Passage, then of the open polar sea, and now of the North Pole, inhibited a scientific understanding of the Arctic. Although explorers touted science in their Arctic campaigns, they ignored it in the field. Even for conscientious explorers, science often conflicted with the demands of geographical discovery. The haste to make distance, for example, generally prevented explorers from staying put long enough to produce a meaningful set of observations. With this in mind, Weyprecht proposed setting up a series of fixed stations in the Arctic maintained by different countries. He introduced his idea at the Association of German Physicians and Naturalists in 1875, and the idea soon gained support within the European scientific community. By 1879, Weyprecht had gained the support of a number of European countries that agreed to participate in an International Polar Year.[16]

In the United States, Weyprecht's plan gained the support of Albert Myer, commander of the Army Signal Corps. For Myer, a high-profile scientific expedition to the Arctic offered an excellent opportunity to raise the Signal Corps' reputation outside of the army. Myer had developed the Signal Corps largely on his own during the Civil War, having convinced Congress of the value of the telegraph for military communications. But with few friends among the army establishment and looming post-war cuts in military appropriations, he found himself on the verge of becoming a "Chief without a Bureau."[17]

As a result, Myer sought allies outside the army, building a base of support among lay and scientific groups. He put his telegraph network and line operators to work in transmitting meteorological information, appealing to civilians and members of the scientific community who were anxious to set up a national weather service. Prior to the war, the Smithsonian Institution had taken the lead in organizing weather data, using telegraphy to collect data from hundreds of volunteer observers around the country. Yet the war had disrupted this network, and a fire at the Smithsonian in 1865 had destroyed valuable meteorological data and equipment. Unwilling to commit the Smithsonian's resources to rebuild and expand the weather service, Henry had welcomed Myer's idea of using military posts and personnel for conducting meteorological work. Both men signed their names to a formal petition in 1869, and by 1870, Congress directed the Secretary of War to initiate a meteorological service, for which Myer and the Signal Corps took responsibility. In this way, Myer used Signal Corps equipment and personnel for civilian and scientific ends, making allies outside the army that helped keep the corps alive.[18]

Although polar exploration seemed far afield from the Signal Corps' responsibilities as a weather service, Myer and his officers formed their own Arctic coterie. In 1869 and 1870, for example, a Signal Corps officer, T. B. Maury, gained attention for his defenses of the theory of the "thermometric gateway" and his criticism of Hall's *Polaris* expedition, both of which appeared in the popular press. Another officer, Captain Henry Howgate, had plans to establish an American polar colony in the high Arctic, and he went so far as to siphon money from the corps to fund his own preliminary expedition to Baffin Land in 1877. Myer became aware of Weyprecht's plan because of his participation in the International Meteorological Congresses and sought to extend American participation in the plan. Although Myer died before he could act on the proposal, William Hazen, his successor as Chief Signal Officer, pursued the issue with the army and Congress, securing funding for an American station in the high Arctic in 1880.[19]

The scientific mission of the expedition presented an excellent opportunity for Myer, and later Hazen, to raise the profile of the corps as a serious scientific agency and thus justify it as a suitable home for the national weather service. The plan certainly drew prominent advocates from the scientific community who supported U.S. participation in the International Polar Year. Here indeed was a plan that seemed to realize Bache's and Henry's earlier vision of systematic Arctic research as the key to a holistic understanding of meteorology and geomagnetism. Not surprisingly, then, Henry wrote letters to Congress in support of the plan and was joined by Matthew Maury, Louis Agassiz, and Elias Loomis, among others.[20]

At the same time, the harrowing duties of Arctic exploration also offered the corps a means of displaying the soldierly qualities of its men to critics who viewed it as a civilian scientific agency in disguise. Although close connections with scientific groups had helped the Signal Corps prove its usefulness and avoid dissolution, it had also distanced its men from the manly qualities associated with soldiering. Army elites already considered the Signal Corps to be too scientific to be soldierly. Testifying before the House Committee on Military Affairs, Commanding General of the Army William T. Sherman stated that corps officers were "no more soldiers than the men of the Smithsonian Institution. But what does a soldier care about the weather? Whether good or bad, he must take it as it comes." Arctic exploration offered the Signal Corps a means of demonstrating the bravery of its men among the hazards of Arctic life. Deprived of traditional opportunities for combat, the unit entered its battle against the Arctic, drawing on the metaphorical union of the Arctic exploration and warfare established by earlier explorers such as Franklin, Kane, and Hayes.[21]

The man chosen to lead the expedition, Adolphus Greely, appeared to embody the combination of scientific and soldierly qualities that formed the twin pillars of the expedition. In the Civil War, Greely had fought extensively and had been twice wounded. Assigned to the Signal Corps after the war, he learned telegraphy and supervised the dramatic expansion of the telegraph network across the western United States. As the corps adjusted to its new role as a national weather service, Greely adapted himself as well, soon becoming the army's top meteorologist. As the press turned its attention to the expedition in 1881, it favorably acknowledged Greely's dual status as soldier and scientist. *Frank Leslie's Illustrated Newspaper*, for example, detailed Greely's war record and then turned to his work as a meteorologist. "Old Probabilities," as it referred to Greely, had become better known for his feats of meteorology (successfully predicting the weather four days in advance of the presidential inauguration) than for any military actions taken during the Civil War. As a result, Greely's persona as an explorer bore less resemblance to George De Long's than to Elisha Kane's thirty years earlier.[22]

During the expedition, Greely reinforced his reputation by dutifully carrying out his scientific mission in accord with the Weyprecht plan. Reaching Lady Franklin Bay with his party on 14 August 1881, he organized his party to collect dozens of daily observations. Yet Greely ran into many of the same pitfalls that had plagued earlier expeditions. With the exception of two Greenlanders whom Greely had hired for the expedition, few in the party had any Arctic experience. The haste of the expedition's departure had prevented an adequate check of the scientific equipment. Solar and radiation thermometers sometimes failed to function. Geomagnetic observations suffered from an inferior dip meter that "materially impaired, if not effectively destroyed, the value of our dip observations." Natural history also suffered under the care of the expedition's naturalist, who failed to adequately label and preserve his specimens.[23]

Perhaps most damaging to the scientific program, however, was Greely's eagerness to pursue geographical discovery, contrary to the spirit of the Weyprecht plan. As the most northerly polar station, Fort Conger provided Greely an excellent base from which to beat the English record for reaching "farthest north," established in 1875 by the Nares expedition. So, starting in the spring of 1882, Greely sent a number of parties out to make distance. On 15 May 1882, a party of three men, James Lockwood, David Brainard, and native Greenlander Thorlip Frederick Christiansen, trekked up the coast of Greenland to latitude 82° 24' north to set a new record, beating the British record by four miles.[24]

Hazen sent a ship to relieve Greely at Fort Conger in 1882 as planned, but thick pack ice in Smith Sound prevented it from reaching the base. A second attempt to relieve the party failed in 1883. Greely had prepared for the possibility that ships might be unable to reach Fort Conger and set off southward along the coast of Grinnell Land in the fall of 1883. Greely expected that these ships would leave large caches of provisions at points further south at prearranged locations. Owing to a combination of ambiguous orders and poor judgment, however, the relief expeditions had left few provisions for the men retreating from Fort Conger. As a result, Greely and his men completed their exhausting exodus depleted of food and deprived of the new supplies needed to get them through the oncoming winter. The lack of relief had exacerbated tensions between Greely and his men. As food dwindled during the winter of 1883–84, his men began to die one by one. When one man, Charles Henry, was caught stealing food repeatedly, Greely ordered his execution. By the time of their rescue in June 1884, all but seven of Greely's party had died of exposure or starvation.[25]

The tragic fate of the party dramatically increased its profile as a news story. In this sense, the story of the Greely expedition followed in the footsteps of the De Long expedition, gaining momentum with the public only after it demonstrated its interest as a disaster story. East Coast newspapers had grown concerned with the story after the failure of the first relief expedition in 1882. The New York Times, in particular, singled out Hazen for his failure to ensure that Greely's party found provisions as they moved south.

Other figures such as the secretary of war, Robert Todd Lincoln, also came under fire from the popular press for the failure of the relief expeditions. Although disaster fueled press coverage of the *Jeannette* and Greely expeditions, the press treated the two differently at first. Whereas editors and reporters had placed members of the *Jeannette* expedition, especially Melville, under scrutiny for their roles in the death of De Long's party, they praised Greely and his party for enduring such hardships, saving their venom for expedition organizers such as Hazen and Lincoln. By comparison, the press largely ignored the men of a second American Arctic station, established at Point Barrow in Alaska, who completed their mission successfully and without mishap.[26]

Shielded from criticism, Greely and his party became celebrities in the first weeks after their rescue. Newspapers widely praised Greely and the other survivors. The navy, eager to show off Schley and the relief expedition, set up an elaborate reception for the explorers in Portsmouth, New Hampshire. As the relief ships glided into Portsmouth harbor for their reception, they were met by large, enthusiastic crowds. The dailies were eager to elevate Greely to the status of hero, emphasizing personal aspects such as his reunion with his wife Henrietta in the closest detail. As she made her way to the deck of Greely's ship, the *New York Times* reported, Henrietta Greely "trembled in every limb, her breath came in gasps, and her whole frame shook with emotion." Seeing his wife, Greely leapt out of his chair to embrace her, uttering a "loud cry, that was more like a gigantic sob smothered." *Frank Leslie's Illustrated Newspaper* devoted its entire front page to an illustration of the scene, the couple kissing next to Greely's chair, which had been knocked over in the enthusiasm of their meeting.[27]

Real or imagined, these stories indicated a shift in the coverage of Greely as an explorer. In 1881, newspapers had focused on his scientific and organizational skills; now no one was calling Greely "Old Probabilities" any more. A commander who had led his men through the horrors of Cape Sabine had to be a man of sterner stuff, of greater passion, than his scientific credentials revealed. Columnists praised his courage and fretted about his health. As thousands of well-wishers descended on Portsmouth for an extravagant reception, Greely directed his first words of thanks to the press. Meeting briefly with an Associated Press reporter, Greely praised him and the other newspapermen for treating him in a "universally kind manner." Shortly thereafter, Greely offered his thanks more concretely, lending the *New York Herald* a private diary of one of the Cape Sabine dead. James Gordon Bennett himself wrote Greely, tendering the *Herald*'s "warmest thanks." So close were the various voices in the presentation of these events that the official account of the reception, written by W. A. McGinley and published by the Government Printing Office, was given to Greely for final approval. "You can now make such changes, alterations, and additions as you desire. You can rewrite it entirely," McGinley wrote, and "give it the shape which best suits you."[28]

Indeed, speakers and members of the press seemed eager to draw distinctions between the Greely and the *Jeannette* expeditions. On 4 August 1884, the town of Portsmouth hosted a parade for Greely and his men in which more than two thousand individuals marched, including the secretary of the navy, the governor of New Hampshire, congressmen, and local groups such as the Portsmouth Fire Department. Later, at the Portsmouth Music Hall, speakers were anxious to point out that the criticism surrounding the expedition did not touch the Greely party. "Nothing dims its record," Senator Eugene Hale of Maine told the audience. "There was no insubordination, no blundering, no losing of the head."[29] *Harper's Weekly* offered a similar assessment of the newly returned explorers: "In the latest story of arctic exploration there are no episodes of human weakness and cowardice to break the force of its showing of human strength and courage.... Pitiable squabbles, mutinies, dissensions, scandals, have come to light as if to show us that man at his best is but a poor creature. We ought all to be thankful that no such pettinesses have come to light to belittle the heroism of the latest arctic explorer, and that there is nothing to indicate that any such have been concealed to be brought to light hereafter."[30]

The eagerness of the press to promote Greely and his men as heroes did not still its new impulse to uncover the scandals of expeditionary life. Even as *Harper's* printed its glowing assessment, the *New York Times* published evidence that Greely's men had resorted to cannibalism in their final winter at Cape Sabine. It further alleged that Greely had conspired with Schley and military officials to cover up the story, as well as the execution of Henry. Indeed, despite the harmonious chorus of military, press, and public, the narrative threatened to fall apart on close inspection. Privately, Greely, Hazen, and Lincoln had been discussing how they could put the best face on the execution of Charles Henry. While Senator Hale told crowds in Portsmouth that Greely's expedition had set a new moral standard, Greely and Hazen considered their options. A few days after the reception, the relief ships delivered the bodies of the deceased Greely party members in sealed coffins to Governor's Island in New York, where they awaited transport to their final burial places. Hazen and Lincoln looked on as the bodies, including that of

Henry, arrived. Despite the fact that Henry had been executed for disobeying orders and endangering the expedition, he was buried the following day with full military honors at the Cypress Hills Cemetery in Brooklyn. The *New York Times*, which wrote about the funeral extensively, labeled him the "unclaimed hero of the Greely arctic expedition." Instead of breaking the mood of heroism and good feeling that had been so carefully cultivated by the military and the press, Hazen and the military let the funeral proceed, crafting a rationale for the funeral when Henry's fate was finally made public. "The funeral which took place with the military honors," Hazen wrote Greely, "can, if necessary, be explained in future by the then apparent propriety of silence."[31]

But Greely and Hazen would have bigger things to worry about. On the same day that Hazen wrote his letter to Greely, the *New York Times* printed a front-page article titled THE HORRORS OF CAPE SABINE, in which it accused the Greely party of murder and cannibalism. According to the *Times*, the rescue party had found unmistakable signs that members of the party had been killed and eaten. Far from the picture painted by Greely and the press of his crew in dignified and peaceful anticipation of death, a picture of men terrified to be the next meal of their companions was revealed. "'Oh,' he shrieked, as the sailors took hold of him to lift him tenderly, 'don't let them shoot me as they did poor Henry. Must I be killed and eaten as Henry was? Don't let them do it. Don't! Don't!'" In addition, Greely and the navy were implicated in attempts to keep the story quiet. The *Times* also offered its readers all of the macabre details, including the fact that many of the bodies had been picked clean, including that of Henry, who had been one of the last to die, which suggested that some or all of the surviving members had participated in the human feast. In an editorial printed on the same day, the *Times* corrected any notion that the Greely expedition could be used as a national model: "[T]he facts hitherto concealed will make the record of the Greely colony—already full of horrors—the most dreadful and repulsive chapter in the long annals of Arctic exploration."[32]

The *New York Times* article unleashed a wave of articles by other newspapers anxious to exploit popular interest in the macabre details of the scandal. Few publishers could have failed to notice how the scandals of the *Jeannette* expedition had stoked interest in the trip and had proved to be a boon for the *New York Herald*. After the *Times* article of 12 August 1884 appeared, newspapers outdid themselves in offering readers lurid details from the expedition, often at the cost of accuracy. The *Rochester Post-Express* convinced relatives of one of Greely's officers, Frederick Kislingbury, to disinter his body in order to verify charges of cannibalism. In exchange for paying for the exhumation, the *Post-Express* gained an exclusive on the story. An autopsy revealed evidence of cannibalism. In Indiana, the story prompted relatives of William Whisler, another victim of the expedition, to exhume his body for examination. Doctors found Whisler's skeleton almost entirely stripped of flesh. Papers that could not offer news of their own relating to the scandal reprinted the accounts of the *Post-Express* and the *Times* for their readers. The *Detroit Free Press*, for example, told its readers that the contents of Kislingbury's stomach suggested that he, too, took part in cannibalism. More lurid stories followed from the *Chicago Tribune* and the *New York World*, both of which followed up on the *Times'* allegations that members of Greely's party had killed each other for food. Eventually, a number of papers began to criticize some of the more spurious allegations. At the same time, military and political officials hatched a theory that the flesh stripped from the bodies had been used by the Greely party as bait to catch shrimp. Together Robert Todd Lincoln and William Chandler, perhaps with the help of the White House, convinced Schley to edit his final report to include this theory. Greely put out his own statement about the charges, denying any knowledge of cannibalism by him or his party.[33]

Yet even as the cannibalism story began to fade, the press continued to raise questions about the conduct of the expedition party. Just as the scandal had begun to die down during the winter of 1884–1885, Lincoln, in one of his final acts in office, declared that private diaries from the expedition should be made public. As a result, the controversy over the expedition roared back to life as the press published diary accounts that depicted the complete breakdown of command and moral order. Newspapers published articles under titles such as "A Sad Arctic Record," "More Arctic Horrors," and "The Deplorable State of Affairs: Insubordination, Grumbling, Thefts, Occasional Drunkenness, and Angry Words and Blows."[34] Such reports reaffirmed the failure of the party to live up to traditional codes of character of Arctic explorers. In particular, the reports proved most unflattering to Greely, revealing that most of his men bitterly disliked him and disparaged his abilities as a commander. One matter that rankled Greely's men was his reluctance to attempt to reach Littleton Island over the floes of Smith Sound. This issue had also become prominent in the press: "Among the questions relating to the management of the party during the last winter which have been in dispute are whether it was possible to cross Smith Sound at any time during the winter to Littleton island, and whether Lieut. Greely refused to allow an attempt to cross to be made."[35] Even Hazen voiced his puzzlement over

Greely's decision to stay on the shores of Grinnell Land. He wrote Greely urging him to explain why he didn't "escape across to Littleton Island."[36]

THE ATTRIBUTES OF MANLINESS

Greely and his supporters fought critics by framing the party's experience of "dying like men" as a test of manly character. When damaging reports surfaced in the press, Greely and his allies pointed to the extraordinary sacrifice and self-control of the men as they starved to death on Cape Sabine. They built their argument on an ideal of manliness that emphasized willpower and self-discipline rather than physical prowess or action. It was an ideal that Elisha Kane's eulogists had expressed quite elaborately and effectively in the late 1850s and that had deep roots in white Protestant culture. In Kane's case, the ideal helped reconcile his manly character with his physical frailty and early death from rheumatic fever.

For Greely and his party, the traits of sacrifice and self-control not only placed them within the venerated company of Kane and other Arctic "martyrs" but also gave meaning to the horrible suffering they endured on Cape Sabine. "All great advances in civilization have been made along the lines of human suffering and death," lectured Theodore Dwight, an AGS member, at a reception for Greely. "This is the secret, the inscrutable law. After the great battle comes progress. Out of individual death spring national well being and life." The principal achievement of the expedition, Dwight continued, was "the exaltation of the moral and intellectual part of man over the physical— the triumph of mind over matter." Within this framework, physical strength played but a small role. Although "the vigor of physical manhood filled our veins to bursting," Greely wrote years later, "we could do nothing."[37]

Greely himself came to embody the ideal of manliness as an attribute of inner will rather than physical exertion. In the same manner as Kane's eulogists, Greely's admirers highlighted the explorer's inner strength of will by emphasizing his physical frailty. "No one who did not talk with Lieutenant Greely would ever take him for a great explorer," wrote one reporter. "His appearance is more that of a superior benevolent American schoolmaster." Another reporter observed Greely on vacation in 1885 as someone who was "careworn in appearance, very quiet in his movements and on hot days w[ore] a white suit and protect[ed] himself from the sun with a large umbrella." Greely encouraged such portrayals. Discussing the qualities of the successful Arctic explorer, he told one reporter, "My experience has proven that an exceptional physique is not in any way a necessity."[38]

But Greely's delicate constitution did not transform him into a Kane-like hero because it carried different associations in the 1880s than it would have in the 1850s. In the years separating Kane and Greely, form and physical prowess gained importance as measures of manliness, a subject that I discuss more fully in the following chapter. Greely may have hoped that his poor physique would highlight the iron will that lay beneath, but others interpreted his frailty as evidence of manly decline. What did white suits and large umbrellas have to do with Cape Sabine? For critics, the foppish Greely was less reminiscent of Elisha Kane than of Oscar Wilde, a man who for them personified the deterioration of male character. Wilde had toured the United States in 1882, lecturing on beauty and aestheticism, symbolizing the ideal as much in his colorful dress—his velvet jackets and famous knee breeches—as in his plays and remarks. That Greely's frail, Kane-like constitution made him seem like a male aesthete was something about which even his supporters commented. "He has superb eyes, black, lustrous and soft as a woman's," wrote the *Philadelphia Times*. Greely "is exquisitely neat—almost a dandy—in his attire." Readers of the New Orleans *Daily Picayune* learned that Greely spoke in a "very pretty... tired, truly-invalid voice." Greely's effete traits did not put off the columnists who wrote about him in newspaper society pages. The *Picayune* observed that Greely served as a "noble specimen of honorable manhood... purified by anguish in the most extreme." Other articles pointed out Greely's extraordinary appeal to women as further evidence of his manliness. For those inclined to see Greely favorably, he was both a courageous explorer and a sensitive, even glamorous man of society.[39]

But to his critics, Greely appeared over-civilized as well as degenerate. Greely's association with aestheticism offered critics an opportunity to attack him on the ground of character. For example, a reporter for the *Buffalo Express* critiqued Greely for being at once effete *and* barbaric. Watching Greely hobnob at a Washington dinner party, the reporter remarked that "you would think to look at him he had quaffed society sillabub all his life and had never eaten raw marine or stewed sailor." Even the title of the article, "Lieut. Greely Aping Oscar Wilde," offered a subtle barb: during Wilde's tour in the United States both the *Washington Post* and *Harper's Weekly* had lampooned Wilde as an ape in aesthete's clothing. Far from serving as a progressive model of manliness, the article suggests, Greely seemed to be modeling the degenerate Wilde. Simultaneously a cannibal and an aesthete, Greely is presented as a close relative of both Wilde and our apish ancestors.[40]

Yet Greely's supporters turned this argument against his critics. Even if he cut a figure more akin to Wilde than to Kane, no one could doubt that he had endured terrible hardships in the Arctic. Who among his critics could attest to a similar experience? Supporters argued that critics should first examine their own behavior for evidence of cultural corruption. "For men, with their comfortable, slippered heels on their fenders, and with their faces flushed with having eaten a luxurious and highly spiced dinner," sermonized one Boston pastor, "to tell us, in their magisterial way, how men ought to behave in their last, desperate fight with famishing hunger and icy death,— this seems to me simply disgraceful." George Melville, who had experienced press attacks at first hand, leapt to Greely's defense against the "silly effusions of the Arctic critic who never ventures his dear life nearer to the Arctic circle than can be seen from the window of some tall printing-house south of 50° N. latitude."[41]

To those who viewed Greely as a harbinger of manly degeneration, his defenders argued the opposite: he represented the vitality of American manhood at a time when it had become endangered. "There are men who believe that the world is degenerating in valor and other manly virtues," Theodore Dwight told an audience in New York, "but when we consider the acts of such men as Sir John Franklin, Elisha Kent Kane, and . . . Lieutenant Greely we will repel the unworthy imputation." Greely's actions offered an antidote to a slackness among Americans at home. "So devoted to our routine work," Dwight continued, "so comfortable in our homes, so rapid in our movements, so luxuriously fed and warmed, that it is well for us to be reminded that outside of all this local refinement and artificial civilization, there are hand-to-hand struggles with nature, in which, if men are not strenuous, she will get the upper hand."[42] Greely also entered the fray, framing his actions as expedition leader as a defense against an encroaching decadence back home. He ended his two-volume narrative of the expedition by urging readers to view it and other manly acts "to penetrate the heart of Africa, to perish in the Lena Delta, to die at Sabine, or to attain the Farthest North" as efforts that might stave off the "decadence of that indomitable American spirit."[43]

THE ECLIPSE OF SCIENCE

It is significant that Greely's scientific labors did not figure in this list of manly prophylactics against cultural decline. Greely and his supporters showed ambivalence about the role that science should play in Arctic exploration. Some commentators were eager to forge a close connection between the scientific spirit and the manly explorer. In language that echoed

Kane's eulogies, *The Earnest Worker* declared: "The nineteenth century has her heroes, but the bravest and best of them do not march at the head of armies. Theirs has been the noble task of defying the powers of nature, and of facing death in a more fearful form than the quick, short contest of the battlefield. They have gone with self-forgetful zeal, in the interest of science, to struggle in long and silent agony with the horrid polar cold and the deadly equinoctial heat."[44] Similar attempts to attach science to manly character unfolded in the exploration community. The Reverend Roswell D. Hitchcock, for example, told AGS members that Greely's expedition represented a "gain to science, to faith, to manliness."[45]

On the other hand, Greely remained reluctant to identify himself too closely with men of science. "Being a soldier . . . it was not my part to stand back from any dangers or difficult work that might fall my way," he told a reporter for the *Pall Mall Gazette*. "If I had been simply a scientific man it would, perhaps, have been different." Greely's reticence about declaring himself a man of science, I believe, reflected both professional and cultural concerns. As a Signal Corps officer, he was aware that his bureau's scientific work had made it unpopular with army elites who viewed officers of the corps, to recall the words of Sherman, as "no more soldiers than the men of the Smithsonian Institution." The furor over the Greely expedition had only exacerbated tensions between the corps and the army. Greely's wariness of being seen as more scientist than soldier would help explain the title of his narrative, *Three Years of Arctic Service*, which acknowledged the military command of the expedition without mentioning its scientific mission.[46]

Yet evidence suggests that Greely's caution also reflected ambivalence about the status of science as a manly activity in Gilded Age America. Greely may have avoided labeling himself as a man of science because it reinforced his reputation among the press as a dandy and an aesthete. Whereas Kane and Hayes had actively promoted their association with science in order to bolster their reputation as men, new postwar concerns about overcivilization diminished the manly appeal of scientific explorers. For example, in reviewing the Swedish explorer A. E. Nordenskiöld's successful navigation of the Northeast Passage over Asia, the *New York Times* complained that the explorer's devotion to science had made his account less "dashing" than it could have been. Whereas the press had praised Kane twenty years earlier for combining the qualities of a scientist and a soldier, the *New York Times* now viewed this as a failing in Nordenskiöld: "His very many-sidedness retards him."[47]

Greely's promotion of the expedition's record of research served to highlight the party's manly sacrifice rather than to imbue it with scientific character. For example, at the AGS reception in his honor, Greely boasted that his party had taken daily barometric readings even as they lay starving at Cape Sabine, evidence that "the men did not forget, during the time of their great distress, the duty upon which they had been sent."[48] He went on to discuss the party's scientific work in similar terms. His discussion of the collection of magnetic observations focused on the toil involved in lugging the heavy equipment. Rather than highlight the broader relevancy of the data collected, Greely used science principally to emphasize sacrifice to a higher cause:

> During our movement southward we carried with us a pendulum, a large, heavy brass instrument, weighing ninety pounds or more, with the box containing it. It was necessary to handle it very carefully, and was an encumbrance and source of great annoyance to us, yet it would be of great value and highly prized if brought back. When we were in straits I said to the men: "Any time that you feel it should be dropped, although I don't like the idea and want to get it back, yet I would not have any man risk his life on account of it." Yet I never heard a man of the party say that it should be dropped. But everyone said, "Hold on to it to the last."[49]

Some members of the scientific community praised Greely's field work more for its expression of manly character than for its scientific value. Rebecca Herzig argues that American scientists embraced sacrifice as an attribute of scientific work in hopes of strengthening their connection to manly ideals in the late nineteenth century. This helps explain the enthusiasm in some quarters of the scientific community for Greely's stories of manly sacrifice. When one Coast Survey geodesist published the results of Greely's magnetic survey, he offered little analysis of the data beyond Greely's story of sacrifice. Hauling the heavy instruments "cannot be too highly commended, when we consider that every pound of dead weight carried necessitated leaving behind so much food to sustain the life of the party on their perilous retreat." Considering the value of Arctic exploration, *Science* presented no specific scientific questions of value but relied on the degeneration arguments used by Greely and his supporters. "In days when luxury and comfort chain so many people to the fireside, and when the occasions for heroic action are so rare, it is good for human nature to witness fresh examples of heroism, all the better that these examples are for the sake of advancing science." Daniel Coit Gilman, president of Johns Hopkins

University, made a similar appeal to his students at a special reception for Greely the following year. "Most of those who are engaged in such studies [of abstract science] are fireside philosophers, wonted to the ease of their homes and the comforts of their libraries and laboratories," Gilman lectured, "but there sits upon this platform one who has gone as far as possible from home, and even from civilization, who commanded a Polar Expedition that carried the flag of our country to a point farther North than any flag had ever been carried before."[50]

Yet the Greely expedition marked a turning point for many members of the scientific community who increasingly questioned the value of Arctic exploration as a vehicle for serious research. For the most part, Greely's success in achieving "farthest north" and the disaster on Cape Sabine eclipsed the discussion of scientific research, as well as the broader scientific goals of the International Polar Year. Emil Bessels, a Smithsonian scientist and one-time explorer, reported that Arctic science "has always been subordinated to a desire to reach the north pole. . . . Had not most Arctic expeditions been animated by those dominant motives, the results would have been of far more consequence." At a meeting of a committee convened by the Naval Institute on Arctic exploration, Dr. H. J. Rink, scientist and former governor of Greenland, spoke in similar terms. "The scientific results to be expected from the [discovery of the North Pole] do not seem equivalent to the dangers and expense caused by it." Even at the AGS, where support had been the strongest, officers began to question whether support for Arctic exploration was consistent with the society's broader goal of building a reputation as a serious scientific agency. Convening a committee to deal with the subject, the AGS quietly sent letters to prominent academic scientists asking them "whether any problems in your department of science would be . . . promoted by further exploration toward either pole." Although three of seven respondents praised Arctic exploration in principle, the remaining four voiced serious opposition. William Davis, professor of geography at Harvard University, complained that Arctic explorers received money for their reckless dashes north when important surveys of states such as New York had languished for want of funds. "As to which one is the more important to New Yorkers," Davis observed, "there can be no doubt." Simon Newcomb, professor of mathematics at Johns Hopkins University, responded, "I doubt whether an expedition would add any [research] of great importance to those already made." Similarly, W. T. Sedgewick wrote: "I have long been of the opinion that the scientific gains of polar exploration are not worth the cost of life and treasure. I have no doubt that the value of the example of courage, hardiness, and fearlessness in the face of great peril is considerable—especially to an ease-loving race. But I do *not* believe that *science* demands these periodic battles with hostile nature."[51]

Perhaps the strongest evidence that science had declined as a means of justification for Arctic exploration came from popular audiences, which tended to have more expansive notions of science than did professionals and were thus more willing to consider expeditions "scientific." Yet even among the former, skepticism about the value of Arctic science had grown. The return of the Greely expedition had prompted President Chester Arthur to declare that the "scientific information secured could not compensate for the loss of human life." Writing in *McClure's Magazine*, Robert Hugh Mill expressed a view of science as something explorers only pursued "incidentally." The engine of modern Arctic exploration, he argued, was not science but the "power of pure sentiment in the quest of glory." One man wrote Greely asking, "Has any knowledge been gained that will be of practical value to mankind and has the research of any of these expeditions, from the 'Franklin,' down, proved of much value to science?" Some supporters of Arctic exploration began to openly acknowledge the lack of science involved in the endeavor. For example, the *New York Daily Graphic* observed that "the only raison d'etre for modern Polar enterprise is the desire to get to the Pole. Probably if we ever do get there, we shall not see anything particularly worth seeing."[52]

CONCLUSION

As much as the tragedies that befell the De Long and Greely expeditions were products of contingencies, of bad decisions and bad luck, they unfolded back home in ways shaped by the ideas and institutions of the Gilded Age. On the surface, James Gordon Bennett's decision to sponsor the *Jeannette* expedition followed in the philanthropic tradition of patrons such as Henry Grinnell. Yet Bennett's interests in the expedition as an attention-getter for the *New York Herald* marked a new commercial phase of Arctic exploration, affecting the way the public learned about explorers and their polar voyages. Although the *Herald* initially promoted the De Long expedition's scientific mission in order to justify its quest for the North Pole, it quickly abandoned this strategy once the expedition proved more compelling as a disaster story. Because rival newspapers considered the De Long expedition to be a publicity stunt on the part of the *Herald*, they did not shy away from criticizing members of the De Long party once details of the disaster became known. As a result, De

Long's and Melville's actions became the subject of more critical scrutiny than ever before, easily eclipsing the controversies that had confronted Kane, Hayes, and Hall. Moreover, the De Long tragedy set an important precedent for coverage of the Greely expedition two years later. When newspapers detected the whiff of scandal, they did not hesitate to publish information damaging to Greely or his party. In scandal, then, newspapers found the means of reviving Arctic exploration as a subject of popular interest. Among the audiences back home, Frederick Schwatka observed, "disaster feeds upon itself and starts the ball rolling."[53]

The new focus on tragedy and scandal further marginalized the role science played in expeditionary campaigns. Although the practice of Arctic science on expeditions had always been uneven, it generally played an important role in explorers' campaigns back home. Science had often justified the danger of polar voyages and been the rationale for their extensive coverage in the popular press. Now danger, tragedy, and scandal became staples of the Arctic campaign, ends in and of themselves for the popular press. When audiences thrilled to the terrible details of the *Jeanette* disaster, Bennett quickly abandoned his coverage of the expedition as a scientific mission in order to focus on the calamity. Coverage of the Greely expedition also emphasized scientific themes until the failure of the relief expeditions and the details of the tragedy at Cape Sabine created more compelling lines of coverage such as cannibalism and murder. Successful expeditions, on the other hand, those that completed their research without scandal or loss of life, were doomed to oblivion in the popular press. Few reporters, for example, covered Patrick Ray's expedition to Point Barrow, where he completed his Arctic research and returned with his party intact.

Moreover, the rise of powerful patrons of Arctic voyages such as Bennett made traditional appeals to the scientific community less important. Unlike their predecessors, De Long and Greely spent little time courting scientific elites, secure in the knowledge that their expeditions were already well-funded and under military command. In turn, members of the scientific community grew increasingly skeptical of Arctic expeditions as vehicles of serious research. Although explorers still retained the sympathies of some men of science, they no longer had the ear of scientific elites, as had Kane and Hayes thirty years earlier. Even popular audiences, who had so warmed to Elisha Kane *because* of his status as a man of science, now found other attributes of manly character sufficient grounds for praise. Indeed, explorers who appeared to be too scientific ran the risk of seeming too refined for the rugged work of Arctic exploration.

The shift of patrons and explorers away from science represented not only changes in the patronage and promotion network of Arctic exploration but also a deeper shift in Gilded Age Americans' ideals of manliness. Concerned that the luxuries of urban life had weakened American men, commentators praised De Long and Greely for their struggles against nature. Whereas Americans had held up Kane as embodying American values at their most refined, they now hoped that explorers offered an antidote to the refinement of American urban culture. Thus in Emma De Long's posthumous account George Washington De Long is seen recapitulating this battle against luxury in his boyhood resistance to his mother's overprotectiveness. De Long's mother was "morbidly solicitous for him" and had a "constant purpose to shield him from danger and accident." Greely's effete tastes and appearance, on the other hand, threatened to damage his reputation as an explorer. Yet he and his supporters responded to critics by using the corruption argument against them. In the end, his terrible suffering in the Arctic insulated Greely from charges of being effete and served, in addition, to make him a celebrated figure in American culture.[54]

Although both expeditions revitalized public interest in Arctic exploration, the popularity came at a cost to explorers. Press coverage of the De Long and Greely expeditions had confirmed tragedy and scandal as vital components of the Arctic story. Explorers hoping to establish themselves as national heroes could no longer look to the press to shield their mistakes from public view. Moreover, public interest in Arctic exploration now appeared increasingly prurient rather than uplifting. Whereas eulogists had praised the public's interest in Kane as a measure of our character as a people, public fascination with the De Long and Greely expeditions seemed to prove the opposite—that Americans seemed more interested in scandal than in personal uplift. "Public notice passed the bounds of eager and intelligent sympathy," the *Earnest Worker* declared, "and degenerated as expressed by the press into a hard and brutal and unsparing curiosity." Even as the popular press would continue to help explorers promote themselves as models of manly character, it now proved equally willing to tear them down if it suited their own interests and those it perceived to be held by its readers. As scandal brought about a renaissance in Arctic exploration, it also ensured that explorers would have to endure fierce storms at home as well as in the Arctic.[55]

———— ❋ ————

CHAPTER FIVE

The New Machines

Walter Wellman and Robert Peary

IN 1894 Walter Wellman pushed toward the North Pole driving dogs and steering sledges, much in the same manner as Kane and Hayes had forty years earlier. But he fared no better than his predecessors. The failure of the expedition, as Wellman saw it, lay with its use of outmoded equipment. The steamer that ferried his party to Spitsbergen, in the Svalbard Islands, fell prey to a fierce storm that drove pack ice into the ship and skewered it "as you stick the tines of a fork through an egg-shell." On Spitsbergen he faced more breakdowns. His dogs, which he had purchased in Belgium, did not take well to the climate and rough ice and soon collapsed. Before turning back, Wellman and his men "mercifully shot every one of them." In 1898 he tried once again, with a new ship and new dogs. Yet this time human frailties proved the greatest impediment to travel. One crewman died during the winter. Others could not traverse the fragmented, slushy surface of the pack ice. On top of these problems, Wellman took a bad fall, fracturing his leg so badly that it turned gangrenous. Unable to continue with his mission, he returned home convinced that a better method existed for reaching the North Pole. "It was whilst pushing and pulling the heavy sledges and boats," Wellman wrote, "that the idea first came to me of using an aerial craft in Arctic exploration. Often I looked up into the air and wished we had some means of traveling that royal road, where there were no ice hummocks, no leads of open water, no obstacles to rapid progress."[1]

Wellman found the means to travel the "royal road" eight years later when he returned to the Arctic commanding the colossal 185-foot motorized airship *America*. The airship promised an end to the miserable days of hauling sledges. But from the beginning, it was clear that *America* had other merits, too. Flying machines and other fantastic mechanical devices had become both the symbols and the spectacles of the age. The giant motor-balloon gave Wellman not only a novel means of reaching the North Pole but also a way of attracting the public's attention and distinguishing his expedition from others that had come before it. Thus *America* forced Wellman to reenvision his expedition plans, in the Arctic as well as on his public campaigns. Earlier explorers had struggled with nature in order to confirm themselves as men of character. Wellman now tried to frame these efforts as the echoes of an earlier age. "The time ha[s] come to adopt new methods," he wrote, "to make an effort to substitute modern science for brute force, the motor-driven balloon for the muscles of men and beasts stumbling along like savages."[2]

Other explorers, however, had little reason to abandon such "savage" methods of travel. Wellman's rival, Robert Peary, grew increasingly dependent on Eskimo equipment and techniques on his long sledging expeditions. Native sledges, fur clothing, tents, and igloos proved cheaper, more reliable, and better adapted to the demands of Arctic travel than temperamental motor-balloons. "Airships, motor cars, trained polar bears, etc.," Peary wrote, "are all premature, except as a means of attracting public attention."[3] Peary's disavowal of new machines had other uses. It played well with his audiences back home, who increasingly viewed Arctic exploration as a release from the overcivilized demands of modern American life.

The differences between Wellman and Peary played out as part of broader debate about the meaning of Arctic exploration at the end of the nineteenth century. One view, dominant since the first U.S. Arctic expeditions had sailed in the 1850s, held that polar exploration was a modern impulse that had as its object the discovery and apprehension of the globe. But a new view, ascendant in the 1890s, held that the real value of polar exploration lay in its rejection of the modern world. According to this line of thinking, explorers sailed north not to extend but to escape the reach of civilization, to find a route that returned them, in a symbolic sense, to the original state of nature. This new debate echoed earlier ones. It rekindled the argument between Isaac Hayes and Charles Hall about the merits of modern and primitive methods of exploration. But it also renewed a much older debate that had flared since the eighteenth century about the merits of civilized life. Did human virtues flower best in the hothouse of modern society or in the primeval wilderness of nature?[4]

Yet the debate also reflected issues that were specific to this moment in U.S. history. Wellman and Peary pitched their expeditions to Americans who

artitechnology

increasingly lived in cities and industrial towns, who saw at first hand the mechanical marvels and social ills that followed in the wake of progress. Americans had mixed feelings about these changes. Wellman's decision to fly a motor-balloon to the North Pole thus took on deeper meanings, becoming a referendum on the effect of modern machines on manly character. On one hand, Wellman's interest in airships reflected wide interest in mechanical devices as a feature of exploration. Popular support for the use of such devices in the Arctic grew despite their repeated failures to live up to expectations. On the other, it put off many Americans who viewed the Arctic as the last untamed region in a civilized world, a region that pitted men in an unmediated confrontation with nature. Among this group, the use of modern machines in the Arctic defeated a major purpose of Arctic exploration: to build and reveal the character of the explorers who traveled there. Nowhere was this more apparent than in Peary's campaigns. Whereas Wellman's campaigns used airships to dramatize his status as a man of progress and civilization, Peary used his campaigns to cast himself as a nostalgic figure, a man of the frontier who had escaped the emasculating influences of civilization.[5]

Machines do not arrive with their cultural meanings preassembled. Nothing inherent in the design or manufacture of motor-balloons made them progressive or manly. Wellman worked hard to endow *America* with these attributes, playing to the cultural attitudes of his audience. Yet Wellman's vision of *America* did not go unchallenged. At a time when the concepts of progress and manliness themselves were in flux, spectacular machines could carry other meanings as well. Many middle-class Americans worried that the luxuries and work practices of civilization might be emasculating American men, and this affected their view of mechanical devices. We see the threat posed by machines to manly character most clearly in Peary's promotion of his new expedition ship, the *Roosevelt*. In his heavy-handed attempts to masculinize his ship, Peary revealed his anxiety that critics might use the *Roosevelt* to undermine his own image. That Peary saw the potential in portraying modern machines as unmanly is further revealed in his critique of other mechanical innovations such as Fridtjof Nansen's ship *Fram*. Although the story of Wellman and Peary demonstrates the importance of the machines as symbols of gender, it also illustrates the degree to which these symbols remained in play.[6]

Despite the differences in their visions of Arctic exploration, both Wellman and Peary placed little emphasis on science. The record of research conducted during U.S. Arctic expeditions had always been patchy.

But explorers in the 1850s and 1860s had taken science seriously as an issue, talking about it a great deal in their campaigns and looking to the scientific community as their natural ally. By the 1890s, this had changed. Explorers had found new commercial patrons who were willing to publicize and bankroll their expeditions, making the support of the scientists less important. For its part, the scientific community had cooled to Arctic exploration as its social position had changed. Increasingly professional, scientists had grown wary of helping amateur explorers hurl themselves at the North Pole only to see them return empty-handed. But the divide that had grown up between explorers and scientists not only reflected shifts in the institutional support for exploration. Ideas about exploration had changed too, taking on cultural meanings that sometimes clashed with those of science. The work of science, after all, had become a powerful symbol of the modern world. And the modern world was precisely what explorers like Peary were seeking to escape.

OUT OF THE ASHES OF FAILURE

The disasters of the Greely and De Long expeditions had revived popular interest in Arctic exploration, but they had done so at considerable cost. In less than three years, thirty-nine men had perished on American Arctic expeditions, five times as many as had died on all previous voyages combined. Although the popular press generally focused on the human failures that precipitated these disasters, its coverage implicated equipment as well. In particular, it challenged the traditional design of ships sent into the high Arctic. From this perspective, the failure of Greely's and De Long's expeditions could be seen as failures of transport, the inability of ships to withstand the terrible pressures of pack ice. The records of previous expeditions only confirmed ships as the weakest link in the chain of polar discovery. Before pack ice stove in the hulls of Greely's *Proteus* and De Long's *Jeannette*, it had sent Hall's *Polaris*, Kane's *Advance*, and Franklin's *Erebus* and *Terror* to the bottom.[7]

In part, these failures help explain the growing popular interest in Arctic vessels toward the end of the nineteenth century. By 1900, the American press often framed the "North Polar Problem," as the *North American Review* called it, as a matter that demanded new mechanical, rather than strategic, solutions. The press emphasized the progress that European expeditions had made in developing their vessels and equipment. *McClure's Magazine* and the *Philadelphia Ledger* kept their readers up-to-date on the successes and failures of the Russian icebreaker *Ermack* in its effort to reach the North Pole. The *Literary Digest* chronicled the efforts of the Canadian Joseph Bernier to lead an automobile expedition into the high Arctic and, along with the *New York Times*, discussed the merits of various rubber compounds for tires to be used there. The *National Geographic Magazine* followed the plans of the Norwegian explorer Fridtjof Nansen to launch his newly designed ship *Fram* directly into the menacing pack ice of the polar sea and told of Nansen's confidence that *Fram*'s radically sloping bow would prevent it from being crushed by the ice.[8]

Although the new attention paid to Arctic vessels and equipment reflected specific concerns about the failure of earlier expeditions, it also grew out of a general interest in, and identification with, machines as a benchmark of American progress. In other words, mechanizing the polar quest was more than the pet project of engineers and newspaper editors. Its appeal had extensive roots among the American public and can be seen in the deluge of letters received by American explorers and geographical societies from individuals offering mechanical solutions to the Arctic problem. One concerned citizen wrote the American Geographical Society proposing to build a railroad to the North Pole. Others suggested the use of submarines and motorized sledges. Robert Peary even received a design for special artillery that could shoot the explorer to the polar axis. "There was an incredibly large number of persons," Peary reflected later, "who were simply oozing with inventions and schemes."[9]

WALTER WELLMAN

No one embodied this spirit of invention more than Walter Wellman (see fig. 10). In 1879 Wellman started his career as a newspaperman, founding the *Cincinnati Evening Post*. He left the *Post* in 1884 to become the Washington correspondent for the *Chicago Herald* and its successor, the *Chicago Record-Herald*. By 1891 he had began covering stories about exploration, traveling to Bermuda in order to determine Christopher Columbus's exact landfall in the New World. In so doing, he fashioned himself after other globetrotting reporters of the day, such as Nellie Bly and George Kennan. Yet Wellman's transformation into a roving reporter-explorer represented more than a fashion of Gilded Age journalism. As the *New York Herald* had demonstrated with Henry Stanley's 1871 expedition to find David Livingstone, such reporters proved enormously valuable in selling newspapers. James Gordon Bennett, owner of the *New York Herald*, had been quick to recognize that the adventures

FIGURE 10. Walter Wellman. From Wellman, *Aerial Age*, p. 8.

of such reporters often proved more exciting to readers than the stories that they were sent to cover. Thus when the publisher of the *Record-Herald* brought up the improbable idea of sending Wellman on an expedition to the North Pole, its owner, Victor Lawson, readily agreed to fund the venture.[10]

Wellman's expeditions of 1894 and 1898 relied on conventional equipment and techniques of travel, but they revealed an early identification with modern machines. At first glance, Wellman's writings appear to fit comfortably within an older tradition of exploration narratives. His accounts, published in the *Record-Herald* and *Century Magazine*, drew on well-worn metaphors of war to characterize his confrontations with nature. But whereas earlier explorers had used such metaphors to frame their direct encounters with the Arctic, Wellman used them to pit nature against his ship, *Frithjof*. In such passages he appears to use *Frithjof* as a proxy for human beings. As the ship steamed north into heavy pack ice, for example, Wellman wrote, "It is a question which is to triumph, water frozen into ice or water heated into steam—nature on one side, or man on the other."[11]

Even Wellman's descriptions of natural phenomena revealed a mechanical sensibility. His writings rarely offered the evocative, Romantic descriptions of the Arctic that had been so common among his predecessors. Kane and Hayes, for example, had brought the Arctic alive using a variety of gothic and classical allusions. Hayes had described the aurora borealis as a "burning Troy" and "the fires of Vesuvius."[12] The northern lights did inspire Wellman to attempt more expressive descriptions for his readers. Instead of connecting them to images of the past, however, he presented them as a marvelous vision of a technological future: "It was just as if all the steam power in the world had been multiplied a million fold, all of it turned to the generation of electricity, and all this voltaic energy were poured through the lenses of vast searchlights placed in every city, town and village the world round; and then at a preconcerted signal by telegraph, all were set playing and dancing upon the very apex of the heavens."[13] The searchlight metaphor said little about the northern lights. It seems to be the metaphor itself—the vision of the vast mechanical reproduction of the aurora borealis—that took Wellman's breath away.

His rose-colored vision of industrial-age machines contrasted sharply with his dreary experience of Arctic travel using traditional equipment. After he broke his leg on his 1898 expedition, Wellman blamed his failures on "savage" methods of travel. He failed to perceive his injury, or his poor sledging progress, to be the result of his inexperience. Instead of setting out in the spring, when the pack ice was stronger, he waited until summer, when warmer temperatures made traveling more comfortable but rendered the ice rotten and ill-suited for sledge travel. Already inclined to view modern means of travel as superior to traditional ones, Wellman took from his injury a commitment to find better methods of reaching the North Pole.[14]

If Wellman needed any confirmation of his views, he found it at the Paris Exhibition of 1900, where he covered the balloon races for *McClure's*

Magazine following his return from the Arctic. He interviewed balloonists such as M. Le Comte de la Vaulx about the superior quality of lighter-than-air travel. Wellman, rendered lame by his sledge journey, listened to the dashing La Vaulx as he preached about the civility of ballooning. "At noon you have luncheon with your family," he told Wellman. "At two o'clock you ascend. Fifteen minutes later you are no longer a commonplace denizen of the easy-going town—you are an adventurer into the unknown, an explorer as surely as any who melt in Africa or freeze in the Arctic." Far from distancing the explorer from the sublimity of nature, La Vaulx argued, balloon travel heightened experience. His description of ballooning, dutifully captured by Wellman, bordered on sexual ecstasy: "One falls to thinking that perhaps he has shaken off the material world and all its belongings, has ceased to be physical and become ethereal; then he rouses from this with a feeling of exultation because man's ingenuity has thus enabled him to triumph over nature, to penetrate her mysteries, to ride straight though them, to soar in space like the gods. One cries out for joy, his voice in a roar which startles him; the popping of a champagne cork is like a cannon shot."[15]

With less flair (and libido), other explorers also considered lighter-than-air travel in the Arctic. For more than a century, Europeans and Americans had experimented with balloon flight. The latter half of the nineteenth century witnessed a surge of popular interest in such travel, and its advocates quickly recognized its potential for polar exploration. In 1852 Robert Mills, architect of the Washington monument, had urged Elisha Kane to use balloons for reconnaissance in the Arctic, an idea suggested by others as well. Kane had briefly considered the idea before rejecting it as impractical. In the 1870s George De Long had also examined balloons as a means of observation on the *Jeannette* expedition and had gone so far as to discuss the idea with French balloon manufacturers. In the early 1880s John P. Cheyne, an officer in the Royal Navy, had put forward plans to get to the pole by balloon and had lectured in New York to gain support for the idea. Coming so closely on the heels of the *Jeannette* disaster, however, Cheyne's plan had not gone over well with American audiences. "His projects are here regarded as visionary," wrote a Maine man, but "balloon journeys are too uncertain." In 1896 the Swedish engineer Salomon Andrée had put such ideas to the test, transporting a French-made balloon to Spitsbergen, Svalbard, in hopes of floating to the North Pole. Poor winds and the lateness of the season prevented him from making the attempt, but the following year he lifted off to the cheers of his ground crew and floated north, never to be heard from again. Andrée's disappearance, however, did little to dampen growing enthusiasm for an airborne expedition to the North Pole.[16]

For Wellman, Andrée's failure only pointed out the need for more modern equipment. Wellman designed his voyage around the use of a steerable, or "dirigible," balloon. He argued that a vessel equipped with motors and rudders would not be prey to the whims of the wind, as had Andrée's. As a safety precaution, Wellman planned to bring a wireless telegraph aboard, hoping to maintain contact with the outside world during his voyage. Presumably this would also prevent his disappearance in the Arctic in the manner of Andrée and Sir John Franklin. Moreover, telegraph communication promised a speedy transmission of his discovery of the pole to the *Chicago Record-Herald*. In 1905 he approached Victor Lawson with these arguments and convinced him to underwrite the construction of a dirigible that could fly to the North Pole. Lawson agreed, contributing $75,000 to the expedition, which was cumbersomely named "The Wellman Chicago Record-Herald Polar Expedition."[17]

Wellman's plan had mixed success in rousing the interest of the press and the public. Supporters tended to share his optimism that new machines would provide a solution to the "North Polar Problem." Among these enthusiasts, it was less important for the motor-balloon to benefit science than to serve as a symbol of American progress. The *Providence Journal*, for example, paid less attention to the expedition's potential to advance research than to "astonish the less venturesome thousands who gaze open-mouthed at its gyrations from the safe surface of the earth." Imagining Lawson's orders to Wellman to steer a dirigible to the North Pole, the *Intelligencer* of Wheeling, West Virginia reflected, "The North Pole, airship, wireless telegraphy. . . . How much of the world's advancement is compressed in that command!" The expedition made real the "inventions of which our fathers never dreamed." In so doing, the dirigible became a powerful symbol of American progress for Wellman and his admirers in the press, a symbol that easily outweighed practical and scientific concerns about the expedition.[18]

For its part, the *Record-Herald* sought to broaden the expedition's base of popular support by emphasizing its scientific value. This mirrored the strategy of the *New York Herald*, which, twenty-five years earlier, had promoted its *Jeannette* expedition by playing up its commitment to science. From its public announcement of the expedition on 1 January 1906 onward, the *Record-Herald* emphasized the value of the expedition to American science. "All agree," that paper proclaimed, "that the trip will be, under any circumstances, of wide scientific benefit." Yet even this positive slant on the Wellman expedition could not fully conceal the skepticism of the scientific community. The failure of so many Arctic expeditions to accomplish or even attempt serious research had soured relations between scientists and explorers by the turn of the century. Few explorers shared the close rapport with scientific elites enjoyed by mid-nineteenth-century predecessors such as Elisha Kane. In the weeks following its announcement, the *Record-Herald* published interviews with scientists from the University of Chicago and Northwestern University who agreed to speak about the expedition's scientific credibility. Many did not believe Wellman's dirigible capable of reaching the North Pole. John Paul Goode, professor of geography at the University of Chicago, was slightly more upbeat, telling the paper that "Wellman may be able to bring back data which will be extremely valuable," such as systematic observations of currents, weather, flora, and fauna. Readers were left to consider whether these observations could be made from a dirigible soaring high above the ice. The *Record-Herald* published the scientists' comments under the subtitle "Vast Good to Science Sure."[19]

Many newspaper editors around the country reacted with a degree of skepticism. By 1906 they had grown accustomed to sensational expeditions sponsored by rival newspapers. Consequently, many of them viewed the *Record-Herald*'s claims of "vast good to science" with a jaundiced eye. "There evidently is more desire to advertise a newspaper and increase circulation," observed the *Denver Republican*, "than to achieve a scientific victory." The *Cleveland Plain Dealer* came to a similar conclusion. "The expedition would seem to promise results more promising to the advertiser than to the geographer." Eager to put the best face on its national coverage, the *Record-Herald* reprinted the *Plain Dealer*'s critical remarks along with others under the title "Newspapers of Country Applaud Walter Wellman's Coming Venture." Yet these bald attempts at deflecting criticism had little effect. Even Wellman admitted that cynicism about the expedition ran deep: "The prevailing popular belief was that the scheme must be regarded as either foolishly reckless or deliberately dishonest. In this view a considerable part of the press joined."[20]

After the publicity generated about the expedition by the *Record-Herald*'s critics, the events of the trip were anticlimactic. In June 1906, Wellman's crew assembled a large hangar for the dirigible on Spitsbergen. Wellman christened his French-built craft *America* (see fig. 11). Although Adolphus Greely and others had urged him to test his airship before leaving for Spitsbergen, he had not done so. When he tested the engines for the first time, they broke down immediately. "Every time a motor was set running, something smashed. If it were not the screw, it was a section of the steel shafting which

FIGURE 11. *America* leaving the hangar. From Wellman, *Aerial Age*, following p. 136.

carried the power from the motor. If not that, something else. We made repair after repair, change after change, reinforcement after reinforcement, in our machine shops. But all to no avail. In the end the car itself, the chassis, the aerial automobile, began to give way. It was not strong enough for the motors; it could not stand the vibration."[21] Wellman had better success setting up a telegraph station. Eager to capitalize on this minor victory in the midst of a broader defeat, he highlighted the use of the device by sending his first message to President Theodore Roosevelt. "Roosevelt. Washington. Greetings, best wishes by first wireless message ever sent from arctic regions. Wellman." He quickly followed up his message by sending a second telegram to his patron and employer, the *Chicago Record-Herald*.[22]

Leaving a crew to remain for the winter, Wellman returned to Europe to await a second attempt in the summer of 1907. When he returned the following year, he struggled to keep the balloon house from completely collapsing. On 2 September, 1907, Wellman and his crew managed to get *America* aloft. "With a thrill of joy we of the crew felt her moving through the air." The joy did not last long. *America* soon encountered a snow squall that reduced visibility. The wind pressing against the enormous profile of the 185-foot airship was too much for the single forty-horsepower engine to control, and *America* was pushed back toward the mountainous islands of Spitsbergen. After two hours of fighting against the gale, Wellman gave up and landed the motor-balloon on a glacier.

Still not dissuaded, Wellman returned to Spitsbergen once again in 1909. On 15 August he set off for the North Pole. But shortly after *America* lifted off, the ship's equilibrator, a long heavy ballast cable, broke off, sending the dirigible soaring into the air. With no means of regulating its altitude, Wellman brought *America* back to earth. Fortunately he and his party were rescued by the Norwegian sealer *Farm*, which had witnessed the whole event. Soon afterwards, Wellman received word that Frederick Cook had reached the North Pole. This news, more than any mechanical issue, brought *America*'s short Arctic career to an end.[23]

ROBERT PEARY

On the basis of his public comments, rival explorer Robert Peary appeared to share none of Wellman's exuberance about machines (see fig. 12). "Man and the Eskimo dog," he told members of the National Geographic Society in 1907, "are the only two mechanisms capable of meeting all the varying contingencies of Arctic work." Yet he had not started out his career with this view. At Bowdoin College he had excelled in civil engineering, and after graduating he had won a competitive appointment in the U.S. Navy in that capacity. His exceptional talents as an engineer led to more opportunities in the navy. The chief civil engineer of the trans-isthmus canal project selected him to survey Nicaragua for a suitable canal route.[24]

Meanwhile, Peary read a report about the unexplored regions of Greenland that rekindled a childhood interest in Arctic exploration. With some money loaned to him by his mother, Peary secured passage on a ship to Greenland in 1886 in hopes of crossing the interior ice cap. Although his attempt to man-haul sledges across the ice cap failed, it confirmed his interest in mechanical solutions to Arctic travel. His first Greenland expedition inspired him with designs for an alcohol stove, a sledge odometer, a superior snowshoe, and other equipment with which he filled the pages of his diary. Returning home, he earned a membership in the American Association for the Advancement of Science on the basis of his engineering knowledge, after giving a lecture to the association titled "The Engineering Features of the Nicaragua Canal." Yet the life of a naval engineer no longer excited Peary, and as he returned to work for the navy, he grew more eager than ever to return to the Arctic.[25]

In later expeditions, Peary brought many of his new designs to life in addition to adopting the ideas of others. When the Norwegian explorer Fridtjof Nansen successfully crossed southern Greenland in 1886, he dashed

Peary's hopes of completing his original expedition route. Nansen had also accomplished the task by man-hauling sledges. But unlike Peary's party, Nansen's party relied on the use of skis that had been refined by Norwegians in the late nineteenth century. After a rough start, skis proved invaluable to Nansen, enabling him to travel up to seventy kilometers per day and to complete the crossing in only six weeks. Peary now resolved to cross the island at a point farther north, starting from the western shores of Smith Sound and sledging northeast across the inland ice cap. Not only would such a trek prove longer and more arduous than Nansen's, but it presented an opportunity to determine the extent of the northern coastline of Greenland, a matter of centuries' worth of speculation. Peary resolved to bring skis for his party as well. Yet when his party set up its Greenland base camp in 1891, no one on the party was able to master the knack of skiing except for a young Norwegian member, Eivind Astrup.[26]

Although Peary failed to incorporate Norwegian equipment into his second Greenland expedition, he succeeded in adopting the tools and techniques of local Eskimos. The North Greenland expedition of 1891-92 brought him into close contact with Eskimos from whom he gained an appreciation of native equipment and modes of travel. As he set up his base camp, Redcliffe, with his wife and his party, Peary increasingly relied on the assistance of the local "Arctic Highlanders," among whom were probably descendants of the Etah tribe who had helped Kane and his crew survive scurvy and starvation in 1854 and 1855. One of the Eskimo women, Manee, prepared sealskin suits for Peary and his crew that protected them from cold and exposure more than their own woolen clothes did. Soon other local women also produced a variety of animal-hide clothes for the expedition. Although Peary initially sought to man-haul sledges across the ice cap, Eskimos convinced him of the value of having dogs pull the sledges. Yet Peary and his men had little aptitude for the art of driving dogs. Only Matthew Henson, Peary's black servant, showed real talent in working with the animals, he soon excelled at speaking Inupik as well. (Despite difficulties, dogs became increasingly important to Peary's campaigns after 1891.) The Eskimos also taught the party how to craft igloos out of snow. Not only were igloos superior to canvas tents for protection and insulation, but they took up no space on the sledges, leaving more room for food. With Astrup, Peary succeeded in reaching Independence Bay on the east coast of Greenland at latitude 82° north.[27]

Peary began to emphasize his use of Eskimo artifacts in his popular campaign back home. "The traveler who goes upon the ice-cap without fur

FIGURE 12. Robert Peary. From Fitzhugh Green, *Peary: The Man Who Refused to Fail* (New York: G. P. Putnam's Sons, 1926), frontispiece.

clothing," he wrote at the turn of the century, "does so either from ignorance or because he is reckless." After his 1891–92 expedition, he purchased his expedition clothing from Eskimos rather than American tailors. Justifying this choice to the readers of *Cosmopolitan*, one member of Peary's party argued that Eskimo women "sew[ed] a seam as fine as any machine could make." Peary's narratives included long passages about Eskimo life and equipment. On his lecturing tours, he prominently displayed pieces of Eskimo equipment such as sledges, harpoons, and skins on the stage. He came on stage wearing furs, often followed by Matthew Henson, who led a pack of Eskimo dogs in their native Greenland harnesses. By contrast, he emphasized the limits of civilized equipment in Arctic exploration: "Sooner or later—and usually sooner—any machine will fall down in polar work," he later wrote, "and when it does so it is simply a mass of old junk which neither men nor dogs can eat, and which cannot even be burned to cook a pot of tea."[28]

In his willingness to adopt "primitive" modes of travel Peary stood in contrast not only to Wellman but also to the popular campaigns of his predecessors. In the 1850s and 1860s Elisha Kane and Isaac Hayes had recognized Eskimos' skills and modes of travel, adopting the use of dog-hauled sledges and sealskin suits. Yet in their eagerness to highlight their own perseverance and pluck, their lectures and writings diminished the critical role that natives played in the survival of their parties. Similar misrepresentations persisted into the 1880s. From 1878 to 1879, Frederick Schwatka and his party completed a three-thousand-mile sledge journey to King William's Land and back in order to recover records of the Franklin expedition. Schwatka's expedition relied on the labor and intelligence of local Eskimos who served as guides, hunters, drivers, and tailors and furnished the party with skin suits, dog teams, igloos, and fresh game. Such details received little attention at home, where the American Geographical Society fêted Schwatka for his tough Anglo-Saxon spirit. Astonishingly, Isaac Hayes viewed Schwatka's long journey with the Eskimos as evidence of the civilized man's triumph over his savage brethren: Schwatka "has matched himself with the savage man, and, by sheer intellectual force, beaten the savage on his own ground, not alone in endurance, but in sagacity in tracking the migratory deer, upon which his success and life depended. An unerring instinct guided him to his goal; and in the pursuit thereof, he has furnished us a conspicuous example of American pluck and energy."[29]

Not all explorers, however, diminished the role of their native companions. Already, in the 1840s and 1850s, the explorer John Rae of the Hudson's Bay Company had understood the utility of Eskimo practices of hunting and travel and had employed them to make extraordinary journeys across the northern reaches of British North America. Yet Rae's travels never captured the attention of popular audiences in Britain and America, who seemed to prefer the peril of ill-adapted expeditions by Sir John Franklin and others. Charles Hall had fared better in the 1860s, generating publicity for his searches for Franklin by placing Eskimos, figuratively and literally, in the foreground of his lectures and writings.[30]

Yet both of these campaigns stood in contrast to Peary's. Rae and Hall tended to emphasize the civilized qualities of their Eskimo companions. This may have reflected the explorers' genuine sympathy for native modes of life; it also proved essential to their credibility with popular audiences. Except for the relics that they procured, Rae's and Hall's reports on the fate of Franklin depended entirely on Eskimo testimony. For this testimony to be seen as credible, both explorers first had to establish the honesty and civility of their native witnesses. By comparison, Peary generally praised the Eskimos because of, rather than in spite of, their so-called primitive traits. This extended to Peary's discussion of native equipment as well. In terms of his popular campaign, Eskimo implements became important insofar as they demonstrated the limits of civilized devices and modes of travel.

SAVAGE VOGUE

Peary's willingness to praise Eskimos and their equipment *as primitive* reflected wide interest in aboriginal tribes and frontier life in Gilded Age America. As we saw in chapter 3, popular audiences had long been fascinated by Indians, the archetypal savage for most Euro-Americans, and representations of Indians pervaded literature, travel accounts, art, and popular advertising. Initially depicted as menacing and cruel, they gradually became more benign figures in the late nineteenth century as white settlers perceived them as less of a threat on the frontier. Widely represented as "children of nature," American Indians often served in literature and art as a symbol for nature. Their steady disappearance from the western territories in the mid-nineteenth century, then, came to represent the end of the frontier itself.[31]

Dismay about the end of the frontier seemed more common among urban whites than among those living on the outback. Whereas 5 percent of Americans were city dwellers in 1790, 20 percent lived in urban centers by 1850. The rise of the cities, with their extraordinary buildings and public works, had thrilled many Americans, especially those who were inclined to see them as symbols of American progress. Robert Mills, the Washington architect who in 1853 had urged Elisha Kane to use manned balloons in the Arctic, also viewed the explosive growth of American cities as "the great phenomenon of the Age."[32]

Yet urbanization raised concerns among many. Since mid-century, cholera, yellow fever, and other epidemics had raged through squalid city neighborhoods, which exhibited disease rates that far exceeded those in rural areas, where contagion had fewer opportunities to spread. For pastors and social critics, these threats of disease were caused by the moral ills of city life. Cut off from the healthy outdoor work of the country, critics warned, office workers were weakened by stale air and cramped quarters. "They quit the fields," the Reverend Edwin Chapin wrote in 1859, "where they might make the grass grow, and increase the abundance of corn, to lean over counters, to stifle at writing desks, and too often, to throw themselves away in the tide of dissipation."[33]

These fears grew more pronounced after the Civil War as more Americans sought their livelihoods in city centers. By 1900, 40 percent of Americans lived and worked in cities, many of them in wretched conditions. Even among middle-class whites, who held higher-paying office jobs, the narrow and often tedious nature of work generated feelings of dissatisfaction and isolation from the natural world. Some believed that urban life had given rise to diseases of overcivilization such as "neurasthenia," a condition signaled by depression and physical weakness. Scholars and social critics feared that city jobs and material luxuries threatened to overcivilize the white race, leaving it vulnerable to the stronger, faster-multiplying immigrant races. Popular magazines considered the threat of "race suicide," while novels such as H. G. Wells's *The Time Machine* presented a disturbing vision of the future in which whites had evolved into a docile and childlike race, preyed upon by their more intelligent, yet more bestial, brethren.[34]

The threat of overcivilization appeared particularly dire to white men. The lure of the city's wealth and pageantry, wrote one pastor, was eroding the good character of young men, threatening to emasculate them. "Commerce is flaunting her sudden successes and intolerable follies of luxury in the eyes of the country, inflaming young men with the aspiration to exchange the honors of health and the independence of home for slavery and effeminization in the town."[35] Clark University psychologist G. Stanley Hall also believed that civilization threatened the manhood of America's youth. Hall linked the condition to unhealthy boyhood practices in modern America such as excessive school work, lack of exercise, and frequent masturbation. Ernest Thompson Seton, founder of the Boy Scouts, brought city boys into the wilderness on the premise that "savage play" would strengthen youths against the ills of civilized life.[36]

From this point of view, encounters with the primitive world came to represent antidotes to the unhealthy ways of the urban life. In an address at the Columbian Exposition of 1893 in Chicago, Frederick Jackson Turner argued that American civilization maintained its strength and innovative spirit because of its interaction with the frontier. Throughout the history of the American republic, he argued, pioneers had grown stronger from their confrontations with the hazards and challenges of western settlement. This pioneer spirit added strength to American culture and provide a means for it to rejuvenate itself. With the closing of the frontier, Turner warned, America had to find new ways of challenging its people so as to avoid the slow decay that seemed to be overtaking Europe. Hall, too, considered "primitive life" a means of restoring the manliness of young boys and encouraged them to play "like savages." In their novels, authors such as Edgar Rice Burroughs, Jack London, and Joseph Conrad examined the power of the savage world to transform the civilized voyagers who encountered it. This theme had become so pervasive in popular literature that one critic dubbed it the "Call-of-the-Wild school of fiction," a body of work held together by a "constant preoccupation regarding the measure of our animalism."[37]

No one better symbolized this new movement than Theodore Roosevelt. Born as a sickly asthmatic to an upper-crust New York family, Roosevelt became a passionate advocate for the "strenuous life." Punctuating his urban career as a politician with long expeditions into the bush to maintain his manly vigor, Roosevelt believed that the savage frontier offered a means by which white men could regain the vitality that they once had as primitive men and use it to inoculate themselves against the decadence and effeminacy of the modern world.[38]

Peary drew on these fears of decline when he framed the failure to reach the North Pole as "a reproach to our civilization and manhood." His willingness to praise the Eskimos for their primitive ways of life thus helped to establish him as man cut from the Rooseveltian cloth. Although it conflicted with the image of the modern explorer espoused by Wellman, it proved exceptionally well-suited to the interests and ideals of his urban East Coast audiences. Here, in states where more than half the population lived in cities, Peary's tough frontiersman image played particularly well.

Nowhere was this truer than in New York City. In 1850 Henry Grinnell and the American Geographical Society (AGS) had established New York City as the center of American efforts to explore the Arctic. By 1900 the AGS had been joined by a dizzying array of geographical and fraternal clubs interested in exploration and frontier life. Some, such as the Arctic Club and the Explorers Club, devoted themselves specifically to exploration. Others, such as the American Alpine Club, the Appalachian Mountain Club, and the Camp Fire Club, promoted a more general interest in outdoor life. Peary belonged to many of these upper-crust clubs and lectured before his fellow members in hopes of raising funds for and interest in his expeditions. In return, he added his rugged prestige to these organizations, offering them a vicarious experience of the savage Arctic.[39]

Eager to present the thrills of primitive life to their members, clubs organized activities that sometimes presented incongruous images. Canadian Camp, for example, offered a semiannual banquet at which members paid for the opportunity to sample unusual game. Dressed in tuxedos and evening gowns, club members supped on leopard, monkey, and other exotic meats, while they listened to explorers and big-game hunters recount their adventures. After presenting the American Museum of Natural History with specimens of a new species of Arctic mouse, *Copex smilo fantasticus*, Peary had the rest of his mouse collection minced and served up at the 1905 Canadian Club banquet. Such novelties seem to have gone over well. Two years later, club organizers approached Peary again for banquet provisions, this time in hopes of obtaining shipboard coffee left over from his previous expedition.[40]

By far the most important club for Peary was one of his own making, the Peary Arctic Club. The club hoped to reinvigorate Peary's campaign after a series of failed expeditions in the 1890s. After successfully traversing North Greenland in 1892, Peary made a second attempt in 1894 in hopes of extending his trek toward the North Pole. The following year he reached his old record at Independence Bay, but dwindling provisions prevented him from going any further. Anxious to salvage something from his expedition, Peary organized a mission to bring back two large meteorites from Cape York. In 1897 he returned to Greenland again to recover an even larger meteorite. When he returned, Peary lent his meteorites to the American Museum of Natural History, which put them on display. The meteorites made him a friend of Morris K. Jesup, the museum's president, who then agreed to became the first president and a leading patron of the Peary Arctic Club.[41]

With substantial support from the Peary Arctic Club and others, Peary organized a new North Pole expedition along different lines. Instead of sledging hundreds of miles across the Greenland ice cap, he planned to take his ship *Windward* northward through Smith Sound, off the northwest coast of Greenland, to the edge of the polar sea. Then, from the top of Grant Land (modern-day Ellesmere Island) or Greenland, he would be able to unload his provisions and equipment hundreds of miles closer to the pole and set off from there with a fresh crew and dogs to sledge across the pack ice that covered the sea. But like so many vessels before her, *Windward* was unable to press through the pack ice of Smith Sound, and Peary was forced to turn back and start his overland journey far to the south, on the eastern shores of Ellesmere Island. Blocked by ice in Smith Sound, he left his ship with his party and made a brutal march northward along the coast of Grant Land in the dead of winter. When he arrived at Greely's abandoned base camp at Fort Conger, Peary's feet were so severely frostbitten that two of his toes snapped off when his boots were removed. He would lose six more toes, ruining any chance for him to make a bid for the North Pole that year. He remained in the Arctic for four more years but only succeeded in covering a short distance on the polar sea. These disappointing expeditions of 1898–1902 convinced Peary that he needed a new, specially designed ship capable of passing through pack ice if he was to reach the polar sea.[42]

THE *ROOSEVELT*

Peary's decision to use a modern vessel to aid his expedition raised new risks at home. By the turn of the century, he had established his image as an Arctic frontiersman, a man whose direct encounters with the "savage" Arctic had strengthened him as a representative of American manhood. His decision in 1904 to develop a new steamer for his expedition seemed to be a change in tack; its construction implied that machines, not muscle, played the key role in reaching the North Pole.

Instead of playing down the role of his new ship, however, Peary promoted it as a rugged vessel that embodied the manly spirit of the expedition. In July 1904 he laid out his plans for the construction of "a massive, powerful steamer" in *Harper's Weekly*. He also spoke of the ship's ruggedness in his speech to the Eighth International Geographical Congress. "She will possess such strength of construction as will permit her to stand this pressure without injury. She will possess such features of bow as will enable her to smash ice in her path, and will contain such engine power as will enable her to force her way through the ice." After framing the ship as a participant in his manly quest, Peary then exhorted members of the International Congress to see Arctic exploration as a test of manhood rather than a test of machines. No matter which nation succeeded in sending an explorer to the North Pole, he told them, "be proud because we are of one blood—the man blood."[43]

In February 1905, as the new ship took shape at the McKay and Dix shipyard in Bucksport, Maine, Peary deliberated about the name of the new vessel with his principal patron, Morris K. Jesup. After first suggesting that it carry the name of Jesup himself, the two men agreed to christen the vessel *Roosevelt* after the man who most embodied the new muscular ethos of manliness in general and Peary's expeditions in particular. In choosing this name Peary found an effective means of distancing his sophisticated new ship from other machines that symbolized the modern world. Press reports tended to accept Peary's vision of the *Roosevelt* as a tough, rather than technically sophisticated, piece of expeditionary equipment. The *Marine Review* of Cleveland, Ohio, adopted Peary's aggressive rhetoric about the *Roosevelt*, "whose mission is to drive into, break down, and force away the ice-fields in front." The *Brooklyn Eagle* published a cartoon of the *Roosevelt* steaming into the Arctic, with Peary confidently standing on deck and the figure of Theodore Roosevelt, clad in the uniform of the Rough Rider and wielding a sword, attached to the prow below. In front of the figures of Peary and Roosevelt, polar bears on ice floes scattered in terror (see figs. 13 and 14). The *Literary Digest* republished the cartoon in its own article about the *Roosevelt*, which also contained a detailed sail plan and hull profile of the new ship. The *Digest* echoed the *Marine Review*'s praise for the ship, calling it "the finest vessel in the history of Arctic exploration" and "the most perfect of its kind."[44]

Battered by the pack ice of Smith Sound in 1905, the *Roosevelt* still succeeded in carrying Peary and his party to the northern coast of Grant Land at latitude 82° north. Yet even from this forward position, Peary was unable to reach the North Pole. The conditions of the pack ice of the polar sea made sledge travel extraordinarily difficult. Across the thousand-mile reach of the polar basin, wind and currents pushed the pack ice unpredictably, compacting floes together into mountainous hummocks or pulling them apart to produce mile-wide leads of open water. Struggling northward with his teams of sledges, Peary claimed a new record of latitude 87° 6' north but then had to turn back. He returned to America deeply disappointed in his failure. Anxious to put the best face on the results of the expedition, Peary and his supporters trumpeted his new "farthest north."[45]

They also turned to the *Roosevelt* for evidence of the expedition's success. As the National Geographic Society prepared to honor Peary for this new record, he corresponded with Gilbert Grosvenor, the powerful new editor of the *National Geographic Magazine*, about ways of emphasizing the accomplishments of the *Roosevelt* at the public ceremony. Grosvenor was an important contact who had proved himself to be a master at the marketing of exploration. During his tenure as editor, the number of subscribers to the *National Geographic Magazine* would grow from 900 to 1,900,000.[46] But celebrating the accomplishments of the *Roosevelt* was a difficult matter. Although the ship had carried Peary to an extraordinary latitude of 82° north, it did not come close to "farthest north" for an Arctic vessel.

That honor was held by Fridtjof Nansen's ship *Fram*. In 1895 *Fram* had come within 350 miles of the North Pole. Like the *Roosevelt*, *Fram* had been designed especially for the unique conditions presented by voyages to the high Arctic. Nansen had built it with sloping sides and rounded bilges "completely smooth, rather like an egg cut in half." The design prevented ice floes from getting a purchase on the ship. When squeezed by the ice, *Fram* popped

FIGURE 13. Theodore Roosevelt and Robert Peary aboard the *Roosevelt*. "Roosevelt and Peary," *Literary Digest*, 18 September 1909, 418.

FIGURE 14. "When the 'Roosevelt' Reaches the North Pole," *Literary Digest*, 27 May 1905, 780.

up above the floes, preventing it from being crushed. Convinced that a drift current moved northward from the coast of Siberia, Nansen sailed *Fram* into the Laptev Sea and then deliberately allowed his ship to be beset by ice. For three years it drifted slowly northwest with the pack ice, reaching latitude 85° north and allowing Nansen to reach a new record of latitude 86° 16' north.[47]

As they considered the promotion of the *Roosevelt*, Grosvenor struck upon a way for Peary to undermine *Fram* while playing up the *Roosevelt* as a symbol of manly virtues. The latter had reached the northern shores of Grant Land on her own power. By contrast, *Fram* had reached her most northerly position by drifting with the ice. Grosvenor urged Peary to emphasize this point, contrasting "the Fighting Ship 'The *Roosevelt*' as against the passive qualities of 'The *Fram*.'" Peary included the suggestion in his remarks at the NGS banquet. Any dinner guests who might have missed the symbolic connections between "the fighting *Roosevelt*," Theodore Roosevelt, and modern manhood were soon assisted by the appearance of Roosevelt himself. Interrupting a speech in the banquet hall, he strode to the podium and gave a lecture praising Peary for embodying "the great fighting virtues" of American culture. Roosevelt's remarks reinforced the explorer's image as an Arctic frontiersman, an American ho demonstrated his manliness in an increasingly effeminate culture. "Civilized people usually live under conditions of life so easy," Roosevelt told the banquet party, "that there is a certain tendency to atrophy of the hardier virtues. And it is a relief to pay signal honor to a man who by his achievements makes it evident that in some of the race, at least, there has been no loss of hardier virtues."[48]

In the years that followed, Peary continued to promote the *Roosevelt* as a symbol of his expedition's manly character. He included the ship in the title of his narrative of the 1905–1906 expedition and published Theodore Roosevelt's remarks in its introduction. In his account of the voyage, he gave full vent to the image of the ship engaged in combat:

> The Roosevelt fought like a gladiator, turning, twisting, straining with all her force, smashing her full weight against the heavy floes whenever we could get room for a rush, and rearing upon them like a steeple-chaser taking a fence. Ah, the thrill and tension of it, the lust of battle, which crowded days of ordinary life into one. The forward rush, the gathering speed and momentum, the crash, the upward heave, the grating snarl of the ice as the steel-shod stem split it as a mason's hammer splits granite, or trod it under, or sent it right and left in whirling fragments, followed by the violent roll, the backward rebound, and then the gathering for another rush, were glorious.... At such times everyone on deck hung with breathless interest on our movement, and as Bartlett and I

clung in the rigging I heard him whisper through teeth clinched from the purely physical tension of the throbbing ship under us: "Give it to 'em Teddy, give it to 'em!"[49]

Peary's controversial expedition of 1908–1909, which is taken up in more fully in chapter 6, also relied on the *Roosevelt*, both in the field and at home, as a symbol of the expedition's manly virtues. Amidst charges by his rivals that he had faked his discovery of the North Pole, Peary presented the *Roosevelt*'s accomplishments as an uncontroversial aspect of his expedition. For example, he contrasted it with more luxurious ships. As the *Roosevelt* sailed into a Canadian port on its way north, Peary compared it with the *Wakiva*, a luxury yacht that had pulled alongside with visitors. "No two ships could be more unlike than these two: one white as snow, her brasswork glittering in the sun, speedy, light as an arrow; the other black, slow, heavy, almost solid as a rock—each built for a special purpose and adapted to that purpose. Mr. Harkness and a party of friends, including several ladies, came on board the Roosevelt, and the dainty dresses of our feminine guests further accentuated the blackness, the strength, and the not over cleanly condition of our ship."[50]

Returning to New York in the fall of 1909, Peary sailed the *Roosevelt* down the Hudson River to join a naval parade held during the Hudson-Fulton Celebration. The celebration provided a perfect backdrop for Peary and the *Roosevelt*. Peary's claim to the discovery of the North Pole coincided neatly with the three hundredth anniversary of Henry Hudson's discovery of the "North River" and the hundredth anniversary of Robert Fulton's "North River Steamboat," arguably a linear ancestor of the steamship *Roosevelt*. Moreover, the celebration took place in full view of Peary's most important audience, urban New Yorkers, at a time when he most needed their support. But the *Roosevelt*'s steering gear broke down fifty miles upriver, and it could not take part in the parade. Peary boarded a tugboat headed for the shore, leaving the *Roosevelt* adrift on the Hudson.[51]

CONCLUSION

In the same manner as their predecessors, Wellman and Peary labored to align themselves and their expeditions with values of the nation. That the two explorers came to promote themselves and their vessels so differently speaks to the deep uncertainties about these values, specifically the value of progress at the beginning of the twentieth century. In general, Wellman's promotion of the airship *America* allowed him to capitalize on one attitude pervasive among white Americans, namely, that modern machines were the harbinger of progress, signaling the entry of the United States into the community of civilized Western nations. By contrast, Peary's wariness of modern Arctic equipment aligned him with a countervailing attitude, also common among urban whites, that held that modern machines served as the handmaidens of luxury and decadence, qualities that seemed to be enfeebling and emasculating the white race.

The ethos guiding each man's representation of expeditionary equipment also influenced his choice of persona as an explorer. Wellman had little formal training, having entered the newspaper business at the age of fourteen. Yet as he promoted his voyaging with *America* as a civilized alternative to "men and beasts stumbling along as savages," he took on the role of the cultivated explorer. Whereas images published in Wellman's earlier narratives present him as a man of action and rugged constitution, later photos from his *America* narrative show him dressed in cravat and top hat.

In the manner of earlier explorers, Wellman turned to science as a way of emphasizing his sophistication. Lacking a scientific degree or credibility among men of science, Wellman still encouraged his readers to see his motor-balloon voyages as occasions for serious research. In his narrative of the *America* expeditions, he called his polar base camp "a scientific village in the Arctic" and wrote of "Scientific achievement [as] the purpose and moving spirit." Such statements not only portrayed Wellman as a civilized explorer but also distanced him from the grisly images of cannibalism and starvation that had deluged American readers after the De Long and Greeley expeditions of the 1880s. Lighter-than-air vessels promised both to end barbaric modes of travel and to prevent the barbaric behaviors that had been seared into public consciousness.[52]

Peary played down his own education and scientific training, shedding the uniform of the naval engineer to don the furs of the Arctic frontiersman. As writers and social critics warned Americans about the threats posed by an overly civilized urban culture, Peary "went native." This not only made it easier for him to promote his use of Eskimo equipment but also suggested a way of concealing his use of a modern steamship by making it into a symbol of the expedition's manly virtues. Imbued with its own muscular persona, the *Roosevelt* came to represent the ethos of the expedition, not merely a device for making geographical exploration easier. In this way, Peary stayed true to the "Call-of-the-Wild" culture of his most important supporters.

Peary's masculine personification of the *Roosevelt* also gave him a method by which to criticize the equipment of other explorers. Nansen's expedition represented a radical departure from previous missions. Instead of fighting the motions of the ice, *Fram* relied on them to bring it closer to the North Pole than any other ship. It returned from its three-year voyage in the polar sea with its hull intact and its crew safe, by any measure an extraordinary accomplishment. Yet only by emphasizing *Fram*'s "passive" behavior in 1907 could Peary find a way of celebrating the *Roosevelt*'s voyages to northern Grant Land. The issue was important enough to Peary that he returned to it again in his final narrative. "Only one vessel, Nansen's *Fram*, had been farther north," he wrote in 1910, "but she had drifted there stern foremost, a plaything of the ice. Again the little black, strenuous *Roosevelt* had proven herself the champion." Peary's comments make clear the degree to which he thought vessels of discovery—*America*, the *Roosevelt*, and *Fram*—not only carried the freight of exploration but also the weight of cultural symbols. Making these vessels abide by human codes of conduct gave Peary a way to dismiss *America* as a modern gimmick and to upbraid *Fram* as an effeminate plaything. Borrowing a metaphor used by explorers themselves, ships and aircraft became new weapons in explorers' war against the Arctic, a way to join the battle with nature but more useful still in fending off their human rivals.[53]

———— * ————

CHAPTER SIX

Savage Campaigns

Robert Peary and Frederick Cook

IN 1907 Frederick Cook left New York without fanfare, sailing into the Arctic as the private guest of the gambler and big-game hunter John Bradley. When he returned two years later, he was a national hero, believed by many Americans to have reached the North Pole. One hundred thousand New Yorkers lined the parade route to greet Cook, a native son of Brooklyn, as he made his way from the harbor (see fig. 15). Although his claim had been challenged by rival explorer Robert Peary two weeks earlier, this fact did not appear to diminish the crowds. If anything, Peary's charge that Cook had faked his polar trek only galvanized support for Cook. "We Believe in You," read the banner above the giant triumphal arch under which he passed in procession. Cook's support was not limited to New Yorkers. Across the country, men and women expressed their conviction that Cook had reached the top of the world. In Cincinnati, the Anti-Tuberculosis League made him an honorary member. In Kingfisher, Oklahoma, a soda fountain worker designed a new ice cream soda in his honor. A Montana mine owner led his men in cheers for Cook in which the explorer's support reached "down to the depths of 2800 feet." Almost unanimously, Cook supporters acknowledged that his graciousness toward Peary in the face of the latter's public attacks had brought them over to Cook's side. Many declared that Cook's conduct, more than any other piece of evidence, convinced them he was telling the truth about his North Pole trek. From the beginning of the controversy, then, character played an important and perhaps critical role in the judgment of explorers and their claims of discovery.[1]

FIGURE 15. Frederick Cook and his wife, cheered by New Yorkers after his return from the Arctic. From *Literary Digest*, 2 October 1909, 513.

The issue of character, long an important feature of U.S. campaigns in the Arctic, dominated public discourse about Cook and Peary. This is because the explorers lacked other forms of reliable evidence that could resolve their dispute. The North Pole offered little in the way of unique objects or geography that could be used to confirm explorers' accounts. Nor did the testimony of Cook's and Peary's companions, most of whom were Eskimos and none of whom were white, do much to convince the largely white middle-class audiences who became absorbed with the controversy. As a result, the press and the public gave greater scrutiny to the ways in which Cook and Peary comported themselves at home, searching for truthfulness in their actions, temperament, and demeanor.[2]

Ultimately, investigation of Cook's and Peary's character only clouded this issue of who reached the North Pole first. Both men failed to live up to the public's expectations. Encouraged by the good manners that Cook exhibited toward Peary, Americans defended him even when serious questions remained about his explanation of events. By contrast, Peary's attacks on Cook soured many Americans on Peary, though he had twenty years' experience as an Arctic explorer. As the controversy progressed, however,

Americans perceptions of each explorer fluctuated, and Cook soon found himself on the defensive for exhibiting many of the same character flaws as his rival.

Yet the actions of Cook and Peary do not entirely explain the public's shifting perceptions. The two explorers had a difficult time living up to the public's expectations because these expectations were themselves in flux. The rival campaigns of Wellman and Peary, discussed in chapter 5, revealed split thinking about explorers in the 1890s. While some praised explorers as the vanguards of civilization, others valued them as figures of the past, sailing into the Arctic in order to escape civilization and its pitfalls. To complicate matters, individuals frequently held *both* views simultaneously. Dozens of letters, articles, and editorials portray Cook as a man who balanced the best traits of the gentleman and the frontiersman. He became a role model ideally suited for the age and for his white middle-class target audiences: a man who had heeded the warnings of Theodore Roosevelt, G. Stanley Hall, and others and entered the wild in order to rejuvenate himself. By contrast, Peary's public attacks on his rival eroded his support among Americans, who began to recast him as a man who combined the worst traits of the city and the frontier: an Arctic robber baron who had tried to monopolize access to the North Pole as well as an uncouth brute whose competitive spirit had clouded his sense of fair play and reason.

The issue of character did little to resolve the Cook-Peary controversy, but it made one thing clear: explorers occupied an increasingly unstable niche in American culture. "Who discovered the North Pole?" was the question on many lips in the fall of 1909. Yet formulating an answer focused one's attention on another question: Which world, savage or civilized, did explorers ultimately represent? No longer bound by close ties to the scientific community, explorers could not call on scientists to defend them, to approve their missions, and to vouch for their credibility. Instead, they looked to powerful friends in the press who patronized and publicized their campaigns. But the press was a dangerous ally. As newspapers had entered the business of Arctic exploration, explorers had begun to seem more like corporate mascots than go-it-alone adventurers. Cook and Peary did not look to newspapers to fund their expeditions but reaped great rewards by selling their stories to them when they returned home. Such deals prompted rival newspapers to take sides in their dispute, leading to volleys of negative reporting. Moreover, lucrative deals struck with the press put off the reading public, who began to view Arctic expeditions as the means for explorers to line their pockets rather than to prove their character. In short, the Cook-Peary controversy brought the financing of Arctic exploration into the light, eroding the image of the selfless explorer by tying the issue of money to character.[3]

COMMERCE AND ARCTIC CAMPAIGNS

When the United States first fielded its expeditions to the Arctic in the 1850s, commerce had been an issue freely talked about in Congress and the press. Explorers and their supporters commonly used commercial arguments to justify the danger and expense of expeditions, promoting them in concert with what they saw as loftier goals such as science and geographical discovery. In 1850 the rescue of Sir John Franklin struck humanitarian and nationalistic chords in Congress. Still, supporters of a U.S. expedition thought it important to gain further support by demonstrating its commercial benefits to the nation. "The expedition . . . will go forth," assured Florida senator David Yulee, "with great hope of benefit to the commercial interests of the country." Of all the possibilities, new whaling grounds and trade routes to Asia offered the greatest chances of being realized. The American whale fishery reached the height of its profitability just as the press and the public became absorbed in the search for Sir John Franklin. The most promising new whaling grounds, in the Bering Strait, Hudson Bay, and Cumberland Inlet, bordered regions central to the Franklin search. As for new trade routes, broad acceptance of the theory of an open polar sea in the 1850s revived hopes not only of finding Franklin but of discovering a viable northwest passage to Asia. These arguments were both plausible and timely, portraying new fisheries and trade routes as potential windfalls of geographical discovery.[4]

That promoters continued to use such arguments even when they ceased to be plausible or timely, however, suggests that there were other forces at work in the use of the rhetoric of commercial utility. For example, expedition promoters used the discovery of new whaling grounds as an argument for exploration though it flew in the face of established facts. None of the voyages undertaken by De Haven, Kane, Hayes, or Hall found significant pods of whales. Nor, for that matter, did any of these voyages find a shorter passage to Asia. By the late 1850s, it became clear to careful observers that the value of Arctic expeditions did not lie in commerce. "Were we to calculate the results of these expeditions according to commercial rules," observed the *North American Review* in 1857, "we should be constrained to acknowledge an almost entire failure." Yet such admissions of failure were rare. They did not dampen the enthusiasm of expedition promoters who continued to promise a commercial bounty from Arctic exploration until the early 1880s.[5]

Attitudes about commerce played an important role in Arctic campaigns. If discussion of whaling grounds and trade routes reflected specific economic conditions in the mid-nineteenth century, it also reflected the important role of commerce within the value system of many middle-class whites. Joan Rubin and Thomas Cochran argue that unique social and demographic conditions in the early republic paved the way for nineteenth-century commercial culture. Less tied to titles and estates than they were in Europe, white men proved more willing to migrate in search of their livelihoods. As they established new communities, they found new means of social evaluation such as character and commercial success. The industrial revolution only underscored these new social codes and reaffirmed the role of commerce as a powerful engine of American progress. By the 1880s, businesses sought to generate greater consumption of their products by increasingly extensive and extravagant advertising campaigns. As a result, middle-class Americans found themselves awash in advertising for new merchandise. In the pages of newspapers, magazines, and mail-order catalogs, and in the aisles of mammoth department stores, Gilded Age Americans found encouragement to express their "pecuniary values."[6]

Ironically, the rise of mass consumerism probably contributed to the decline of commercial themes in Arctic campaigns in the 1890s. The explosive rise of consumer culture in the late nineteenth century may have persuaded many Americans that they could buy their way to happiness, but it troubled others who began to view unfettered capitalism as a threat to the moral and cultural life of the nation. In particular, Arctic campaigns of the 1890s and early 1900s reflected this new cultural anxiety about the over commercialization of American life. Whereas earlier supporters of Arctic exploration had bent over backwards to promote the practical payoffs, Gilded Age promoters tried to play down the connections between exploration and commerce. Indeed, they defended Arctic exploration as an *anticommercial* pursuit with little or no practical value. The discovery of the North Pole, they argued, had value to the nation because of, not in spite of, its uselessness to industry or commerce. This rhetoric extended to the explorers themselves. After returning from his harrowing expedition in the Arctic, Adolphus Greely wrote about the new generation of explorers who "were lured by no hope of gain, influenced by no spirit of conquest."[7] Walter Wellman promoted his forays into the Arctic according to this anticommercial rubric: "In this plodding commercial age, this day of humdrum money grubbing and of the routine though admirable round of quiet duty doing, it is a good thing, I think, for the few of us who can to leave the beaten track, fare forth into strange fields, and strive mightily to do things which are exceedingly difficult and dangerous and the more fascinating because they are difficult and dangerous."[8]

These anticommercial sentiments concealed the growing exploitation of Arctic exploration as a way of making money. Although explorers and their patrons no longer held out much hope of finding a northwest passage or hidden whale fisheries, they had developed efficient ways of reaping profit from Arctic expeditions. In the 1850s, no one predicted the extraordinary popularity achieved by Elisha Kane. The success of his published narratives, which sold 150,000 copies after his death, took publishers by surprise. By the 1870s, they were better prepared. *New York Herald* publisher James Gordon Bennett had worked out a system for making exploration pay. By funding his own expeditions in which reporters took on the role of explorers, Bennett generated exclusive news stories about discovery. Henry Stanley's search for David Livingstone had boosted the *Herald*'s circulation and made Bennett famous. Later he successfully exploited the tragedy of the *Jeannette* expedition for similar ends. Other newspapers, such as the *New York World* and the *Chicago Record Herald*, soon followed suit, fielding their own globe-trotting reporters to generate sales.

Even explorers who remained unaffiliated with Bennett or other publishers, such as Adolphus Greely, found new opportunities in the explosion of newspapers and magazines in the late nineteenth century. Although the popular press maligned Greely in the months after his return, it also made him famous. Magazines such as *Harper's Monthly*, the *Century*, and the *North American Review* courted Greely for articles about his terrible voyage. Lecture organizers, one of whom promised him $40,000 for a season on the lecture circuit, also sought him out. Greely later dismissed claims that he had made a profit from his notoriety as an explorer: "Returning penniless, I refused to appear on the public platform, although a small fortune was offered." But Greely was less reluctant than he claimed. In the years following his return from the Arctic, he made extensive lecture tours and wrote articles for a wide variety of popular magazines such as *McClure's*, the *North American Review, Ladies' Home Journal, Youth Companion, Current*, the *Chautauquan*, and the *National Geographic*. He also continued to publish books about Arctic exploration well into the twentieth century. The lesser-known members of his expedition also made money from their association with the expedition, albeit in more modest ways, by performing in plays, giving lectures, and appearing in dime store museums.[9]

Robert Peary best illustrates the growing chasm between explorers' on-stage roles as selfless voyagers and their behind-the-scenes roles as entrepreneurs. By the turn of the century, for example, Peary had become a symbol of the new rustic explorer. His image as a rugged frontiersman not only appealed to audiences anxious about overcivilization but distanced him from the "humdrum money grubbing" of modern American life. In his lectures, he emphasized the purity of motive involved in Arctic exploration. "There is no higher, purer field of rivalry," Peary told members of the 1904 International Geographical Conference, "than this Arctic and Antarctic quest."[10] Purity of motive also underscored his remarks to the National Geographic Society in 1907. "The true explorer does his work not for any hopes of reward or honor, but because the thing he has set himself to do is a part of his being, and must be accomplished for the sake of the accomplishment."[11] His advocates pointed to his fixation with reaching the North Pole as evidence of his purity of motive. "How can this be called a sordid age," one author wrote, "when such men give up the best years of their life to the unrewarded perils of arctic toil?" Far from offering financial rewards, one admirer wrote Peary, his expedition represented his "utter effacement of self." Gilbert Grosvenor, editor of the *National Geographic Magazine*, declared that Peary had "given the world a notable example of a brave and modest man" who had pursued exploration in the face of "financial discouragements."[12] In a letter to Peary, the Connecticut novelist Constance Du Bois praised him for offering men an example of manliness that was free from the modern vices of materialism. "I am sometimes discouraged by the sordid greed of our civilization based on commercialism. The glory of manhood seems to have departed. But you, and your ideals, justify it to my mind—and the response from the people, the men and growing boys, as their spirits still ring true to the appeal of noble adventure, is so encouraging that we need not yet doubt the future of America."[13]

Belying these lofty representations, however, Peary had grown increasingly mercenary during his Arctic career. In part, Peary's preoccupation with money grew out of the enormous expense of his expeditions to the North Pole. Although he launched his first expedition over the Greenland ice cap on a shoestring budget, later ones grew more expensive as they became more ambitious. In order to raise money for a second voyage to northern Greenland in 1893, for example, Peary delivered 165 lectures in a little more than three months, a feat that raised $20,000. But lectures alone could not cover the extraordinary expense of Peary's new ship, the *Roosevelt*, which cost $130,000 when equipped and provisioned. For this he relied on Morris

Jesup and the wealthy members of the Peary Arctic Club. For his part, Peary established a reputation for honoring his biggest patrons for their generosity. In the 1890s he returned with scores of specimens for Jesup's American Museum of Natural History, including three massive meteorites that he had brought back from Greenland. His donations were not limited to animals and artifacts. When museum anthropologist Franz Boas suggested that Peary bring back an Eskimo from northern Greenland for study, Peary returned with six, four of whom died within a year from tuberculosis.[14] Other patrons received gifts commensurate with their support. Local promoters and lecture organizers often received specimens, and Peary favored his female supporters with blue fox pelts. Larger donors found their names immortalized on the capes and bays of Greenland and Grant Land (Ellesmere Island). Not all of Peary's efforts went to pleasing patrons and covering costs. On his final expedition of 1909, he filled his personal notebook with advertising schemes and other means of capitalizing on his fame as the North Pole's sole discoverer.[15]

FREDERICK COOK

The sudden appearance of Frederick Cook as a rival claimant to the discovery of the North Pole cut Peary's business plans short. Cook, a former member of Peary's northern Greenland expedition, sought to make his career as an explorer pay. From a young age, Cook had struggled for financial security. The early death of his father forced Cook and his brothers to become entrepreneurs. He made money in milk delivery and eventually bought a delivery business for himself. Passing the business on to his brothers, he pursued a medical degree in New York City. In 1887 he entered the College of Physicians and Surgeons at Columbia University. Shortly before he passed his final exams, however, his wife died in childbirth. Devastated by her death, Cook volunteered for Peary's expedition as its surgeon. He relished his experience in the Arctic with Peary, and it whetted his appetite for further polar voyaging.[16]

It also led him to appreciate the Eskimos of northern Greenland as a precious, and lucrative, resource. Cook recognized that Eskimos held the key to effective exploration of the Arctic. He marveled at their ability to adapt to their harsh environment. He spent a great deal of time studying their language and customs. He even began to take on their appearance, wearing skin clothing and letting his hair go uncut. Returning home, he also perceived the value of Eskimos in his popular campaign, much as Charles Hall had thirty years earlier. Like Hall, Cook made the Eskimos the focus of his popular lectures. Yet Cook tailored his portrayal of the Eskimos for late-century audiences more likely to be disenchanted with elements of civilized life. Whereas Hall underscored the natural civility of his Eskimo companions, Cook emphasized the distance between them and white explorers, praising them for their undiluted savage ways. "With all our civilization," he told the Kings County Medical Society of New York, "there were few points that we could suggest to them to make them more comfortable in their cold and icy homes." Encouraged by his reception by the society, he made plans to lecture more widely and to publish an account of his expedition.[17]

It was Cook's attempts to capitalize on his expedition experiences, not any faults as an explorer, that led to his first problems with Peary. In the Arctic, Cook had impressed Peary with his stamina, calm temperament, and positive attitude. Confident of Cook's abilities as a leader and explorer, Peary had placed him in charge of his base camp while Peary trekked across the interior ice of Greenland. But back home, Peary feared that Cook's popular accounts of the voyage might compete with his own. Thus when Cook asked Peary for permission to publish his writings, Peary refused. Unable to profit from his Arctic experiences, Cook backed out of a position that he had already accepted as surgeon on Peary's second expedition to northern Greenland. Instead, he organized a private expedition of his own. In 1893 he sailed to western Greenland with a small party of tourists and hunters. Although Cook made no geographical discoveries on his short voyage, he used it to prepare himself for a new lecture campaign afterwards. When he returned to the United States, he brought two Eskimo children, Kahlahkatak and Mikok, Greenlandic dogs, and two barrels full of Eskimo bones that he had disinterred from an ancient grave site. These exotic wards and relics generated a good deal of interest in his lectures.[18]

Encouraged by the publicity, Cook increased the size of his Eskimo retinue. The World's Columbian Exposition in Chicago in 1893 had included an "Eskimo Village" inhabited by sixty Labrador natives. When fair promoters failed to honor their promise to send the villagers back to Labrador, Cook arranged to bring ten of them to New York. There his entourage captured the attention of lecture promoter Major J. B. Pond, who engaged Cook and the Eskimos for a tour through Massachusetts, Connecticut, New York, New Jersey, and Pennsylvania. Cook also gave lectures with his Eskimo party at the Huber Dime Museum and the New York Obstetrical Association. He even entered his dogs at the Westchester Kennel Club Show, where they won three awards. Cook's lecture tour generated interest in his second Greenland expedition, undertaken in 1894, in which he served as a guide to tourists and scientists. The expedition faired badly, however, when the *Miranda* collided with an iceberg off of Labrador and later struck a reef off the coast of Greenland.[19]

The failure of the *Miranda* expedition led to a subtle shift in Cook's modus operandi as an explorer. Until that time he had made ends meet as an Arctic tour guide. After 1894, however, he participated in increasingly dangerous expeditions focused on geographical discovery. In 1897 he sailed to Antarctica as the medical officer of a Belgian expedition to the Antarctic, a harrowing voyage in which Cook's calls for the consumption of raw penguin meat prevented the death of crew members from scurvy. In 1903 he made an unsuccessful attempt to scale Mt. McKinley with a small party. In 1906 he tried again and claimed success when he returned home. He was subsequently honored by the National Geographic Society for his accomplishment.[20]

In 1907 Cook embarked on the most ambitious and controversial expedition of his career: a voyage to the North Pole. The plan started modestly when a wealthy casino owner, John Bradley, asked Cook to be his guest on an Arctic big game hunt. Cook agreed to sail as Bradley's guest. As the date of departure grew near, Cook asked Bradley if he could use the expedition to launch a trek to the North Pole once they reached the Arctic. Bradley agreed, picking up the extra expenses for equipping and provisioning Cook's mission. Bradley's ship sailed into Smith Sound in August 1907 and deposited Cook, along with crewman Rudolph Franke, in Annoatok, on the eastern shore of the sound. In the spring of 1908, Cook, Franke, and a party of Smith Sound Eskimos crossed the sound to Ellesmere Land. Cook set off westward to Axel Heiberg Land. By the time he reached the west coast he had sent Franke and all of the Eskimos' party back except for two Eskimo men, Etukishuk and Ahwelah. When Cook reappeared at Annoatok a year later, in the spring of 1909, he reported reaching the North Pole on 22 April 1908 after a long trek over the polar ice cap. He had to wait until the fall of 1909 to leave Greenland on a Danish ship headed for Copenhagen. While stopped in the Shetland Islands, Cook wired the news of his discovery to the *New York Herald* on 1 September 1909, and the story ran as a front-page headline the following day.[21]

Cook's writings reflected the evolution of his persona as an explorer from Arctic tour guide to rugged discoverer of the North Pole. Articles chronicling his early expeditions played down their dangers. In an 1894 article titled "The Arctic Regions as a Summer Resort," for example, he tried to dispel the menacing image of the Arctic that had been reinforced by the disastrous expeditions of the 1880s. "Each succeeding explorer," he wrote, "has apparently tried to outdo his predecessors as far as harrowing accounts of dangers and difficulties were concerned." Instead, he urged readers to see the Arctic as both safe and exotic, a "new Eldorado of nature's gifts." The article suggests that Cook was appealing to potential passengers for his Arctic voyages. It struck a delicate balance between the image of the Arctic as a frontier, a place of escape from the materialism of civilized life, and an exclusive spa designed to "make life comfortable and enjoyable." On one hand, Cook assured the wealthy and middle-class readers of *Home and Country* that the Arctic was not beyond their reach. On the other, he identified it as a place far enough from civilization to serve as an antidote for its ills. "The pure, exhilarating air and the delightful climate suggest untold benefits to the invalid and rest for the 'world weary.'"[22]

As Cook pursued more dangerous and sensational missions, he abandoned his allusions to upper-class luxuries, representing his expeditions not only as escapes from the civilized world but passages back in time to simpler and more savage days. The most eloquent articulation of this premodern vision came not from Cook but from Robert Dunn, a member of his first Mt. McKinley expedition party: "The true spirit of the explorer is a primordial restlessness. It is spurred by instinct of pre-natal beginnings and a cloudy hereafter. . . . [Explorers are m]en with the masks of civilization torn off, and struggling through magic regions ruled over by the Spirit of the North or the South; human beings tamed by the centuries, then cast out to shift for themselves like the first victims of existence."[23] When Cook emerged from the Arctic in 1909 claiming to have reached the North Pole, his writings and lectures underscored this antimodern vision of exploration. Lecturing about his expedition at Carnegie Hall, for example, Cook portrayed it as a thoroughly primitive and masculine endeavor, echoing Peary's portrayal of his own expeditions: "We used no aeroplanes, no submarines, no motorcycles. We reduced the distance to the simplicity of primitive man. We lived a life as simple as that of Adam, for there is no Eve at the pole to break the frigid silence. Civilized man does not often descend to the scientific limits of the savage. He will live otherwise as long as he can. We had no other way."[24]

Cook's portrayal of his seemingly simple and primitive North Pole odyssey also strengthened his image as a man who explored the Arctic with a pure heart, unsullied by issues of fame or commercial success. The explorer Anthony Fiala concluded from Cook's voyage that, apart from any scientific observations (of which there were few), "there can be no commercial benefit." The editor of *The Discovery of the North Pole,* a hastily written book about Cook and Peary's expeditions, went even further. The purity of the expeditions as quests made matters of usefulness irrelevant. "It matters not," he wrote, "whether we can make up our minds as to the practical usefulness of the North Pole." In his introduction to this book, Adolphus Greely struck a similar note: "The most distinctive feature of polar exploration is not generally recognized, that is, its entire disinterestedness."[25]

[handwritten margin note: disinterestness]

CONTROVERSY AND CHARACTER

Robert Peary rent this veil of disinterestedness when he accused Cook of lying about his trek to the North Pole. A week after Cook cabled news of his discovery to the *New York Herald,* Americans learned that Peary, also on his way back from the Arctic, claimed the discovery for himself. The news turned a big story into a monumental one. Peary's charge that Cook faked his discovery made the story all the more interesting. "Cook's story should not be taken too seriously," he cabled from Labrador. "The Eskimos who accompanied him say that he went no distance north. He did not get out of sight of land." Later he claimed that Cook had handed the public "a gold brick." Peary voiced doubts that had already been raised by many in the press about Cook's claims. Yet coming from Peary, the charge struck many Americans as partisan and self-serving. It flew in the face of his statements about explorers who pursued discovery without "any hopes of reward or honor." By contrast, Cook avoided making any critical statements about Peary's claim publicly. On hearing the news of Peary's discovery, Cook told the press, "I believe him!" and said, "There is glory enough for us all!" His magnanimous remarks heightened his contrast with Peary. Popular support for Cook swelled, whereas many looked on Peary's charges as the desperate claims of a sore loser.[26]

Although the character of Arctic explorers had always been important to the press and the public, it took on added significance in the North Pole controversy because other means of proof remained insufficient or unpersuasive. To some degree, issues of evidence depended on the difficult process of determining the North Pole as a geographical location. Lying near the center of an ice-covered sea, Earth's northern axis possesses no prominent landmarks that can serve to confirm an explorer's passage. Depth soundings served as one means of linking the North Pole to the topography of the ocean floor, but neither Cook or Peary did much to avail themselves of these measurements to defend their claims. Cook made no soundings at all because, he claimed, the equipment was too heavy for his sledges. Peary took sounding equipment but provided only three soundings in his preliminary report. He claimed that he took a sounding close to the North Pole showing a depth of fifteen hundred feet, at which point he ran out of line, rendering his sounding essentially useless. Although both men claimed to have left markers such as flags and cairns at the pole, no one expected these artifacts to remain there long because of the constantly shifting pack ice.[27]

As a result, confirmation of each explorer's claims rested heavily on their written narratives and on the astronomical observations that they used to determine their geographical position. Partisans and pundits argued about the relative merits of each man's account to little effect. Peary claimed that Cook's Eskimo companions reported that they had not traveled more than two days over the polar ice cap before returning to land. Cook's supporters countered that interviews conducted by Peary's men could hardly count as impartial. Even if we assume that Peary's men had not pressured the Eskimos or mistranslated their statements, publishers discounted Eskimos as credible witnesses for either camp. Peary also took issue with the speed at which Cook had crossed the ice cap. Peary's diaries soon revealed, however, that he had made even grander claims as to his rate of sledge travel over the ice cap, far greater than any explorer had ever achieved previously. Both men published photographs of the North Pole, but there were few details that could identify them as being at the top of the world. Indeed, it appears that both men altered or cropped their photographs, omitting features such as shadows or the sun's altitude that could be used to gauge their legitimacy. Yet even if such photographs and astronomical records proved consistent with conditions at the pole, they did not constitute definitive proof. With knowledge of the sun's predicted altitude at latitude 90° north, either explorer could have generated observations and photographs from some other location on the ice cap that appeared to be the North Pole.[28]

Given the difficulties of establishing proof of each explorer's claim on the basis of evidence, the public increasingly assessed each man on the basis of his character. Walter Wellman, having returned from his own expedition, put the point plainly: "There are three ways to test the good faith of one who claims to have been to the pole—first, by his character; second, by his narrative; third by his astronomical observations . . . if character and narrative be impeached, a traveller's alleged astronomical observations are of no value, for the simple reason that, having concocted a story, such a man would not hesitate to concoct astronomical observations to match it—something very easily done."[29] As a result, press reports tended to use evidence from narratives or observations insofar as they illustrated a pattern of truthful or deceitful behavior. Yet the increasingly partisan nature of the controversy, which lined up pundits and publishers on both sides, inhibited reasonable debate about character. "Even men of recognized standing have been thrown off their balance," lamented the *Springfield Republican,* "and have veered violently from one side to the other, with no valid reason in either case."[30]

Within this maelstrom of praise and ridicule, writers, reporters, and private citizens revealed a spectrum of beliefs about manly character. In particular, they exposed the thin border that separated good character from bad, a boundary seen most clearly in the seemingly incommensurable issues that served as the twin pillars of the character debate: first, to what degree had each explorer conducted himself as a gentleman, and second, to what degree had each effectively shed civilization–or at least its vices–to become a true man of the frontier? In this debate traits that operated as positive attributes also had pejorative functions. The actions that revealed Cook's and Peary's good character were often used by critics to show their moral deficits.

GENTLEMEN AND CONFIDENCE SHARPS

Cook's reluctance to criticize Peary made Cook a gentleman in the eyes of many. In letters and telegrams, dozens of Americans praised him for exhibiting a cultured bearing in response to Peary's charges. "I wish to congratulate you on your gentlemanly conduct of controversy," wrote one man. "Most sincere congratulations on your discovery of the Pole," wrote a Maine woman, "and more for your gentlemanly and fine standard of accepting conditions as they are." Although it would be unwise to generalize about popular attitudes from this self-selected group of Cook admirers, they do suggest the important role that character played, not only in raising sympathy for Cook but in strengthening his claims to have reached the North Pole. "Only a man of the greatest character," wrote a Mississippi woman, "would have accomplished what you have and deported himself in the manly way in which you have met the storm since your return." In similar terms, an Oklahoma man wrote Cook that he was "convinced by your manly demeanor that you discovered North Pole as you say." Cook's "dignified reticence and fine magnanimity," wrote a San Francisco man, had persuaded the public and "made them believe that you are a thoroughly truthful, sincere, and high-minded man." Accepting the Medal of Freedom of the City of New York, an honor previously given only to two other men, one alderman spoke of Cook's desire to postpone the ceremony until all the proofs of his discovery were in. It was an action in keeping with "his characteristic manliness."[31]

If Cook appeared to exhibit the best attributes of a modern civilized explorer, the "characteristic manliness" of a gentleman, Peary seemed to display the worst qualities of an overcommercialized society. Reporters admonished Peary for approaching exploration as a business rather than, to use his own words, something to "be accomplished for the sake of the accomplishment." Peary added to this perception by using proprietary language in talking about the Eskimos as "my Eskimos," as well as his use of Smith Sound as his route to the North Pole. To many in the press and the public, his charges against Cook seemed petty and legalistic. The *Los Angeles Express* lampooned him for these attributes in an article called "Scientific Proofs": " 'I will now,' reports our bold explorer, 'proceed (copyright) to give a full account (copyright) of my discovery of the North Pole (copyright). I am a member in good standing (copyright) of the North Pole Discoverers' Trust (Copyrighted in Europe, Asia, Mexico, and the United States. All rights reserved). I obtained its license in due form (copyright) and was given exclusive rights of discovery.' "[32] Cartoonists and writers also made fun of Peary on similar grounds. The preface to one book about the story presents an illustration of Uncle Sam looking at the North Pole, tagged "Peary's Pole," with a sign reading "Private. All explorers are hereby enjoined from discovering this pole by order R. E. Peary Pole Preemptor." Cook generally refrained from speaking against Peary in public, but he did allude to his legalistic charges in a speech before the Arctic Club banquet in New York. Cook asked his guests, "Am I bound to appeal to anybody, to any man, to any body of men for a license to look for the Pole?" Less partisan accounts of Peary also contributed to a view of him as a businessman rather than an explorer. Adolphus Greely's tepid praise for Peary's "mastery of arctic organization and administration" did not help his image as a rugged frontiersman, nor did his observations about Peary's extensive network of wealthy friends back home: "In short, all that money, friends, and experience could command were his." In these ways, serious and satirical portraits framed Peary as a man who had succeeded as an explorer because of bureaucratic thinking and the support of powerful elites but who, in the process, had lost his inner bearing as a gentleman.[33]

As the controversy progressed, Cook's image as an honest gentleman also began to erode. The good manners that he displayed toward Peary could not hide his exploitation of the controversy for financial gain. Shortly after his return from the Arctic, one literary agent wrote to Cook that his feud with Peary only heightened his marketability with publishers. "In my correspondence," he wrote, "I have taken the ground that the Cook-Peary controversy will benefit rather than injure the sale of your works. The book in advance is freely advertised—and sale should be enormous." Cook seems to have taken the message to heart. After receiving $3,000 from the *New York Herald* for his first wire about his discovery, he agreed to publish a serialized account of his expedition in the *Herald* for $25,000. He quickly sold the book rights of his narrative to Harper and Brothers and, within days of his return to America, embarked on an ambitious lecture tour that took him across the country. In a little over a month Cook gave almost thirty lectures in twelve states, from Maryland to Montana.[34]

If the lecture tour added to Cook's celebrity and to his receipts, it also threatened his image as a scrupulous explorer. The *New York Post* urged Cook to stop lecturing before it was too late. "From now on, he ought to be spending all his days and nights in the work of clearing his honor; and every dollar that he takes in henceforward by exploiting his claim when he ought to be removing the cloud on it, will be a dollar gained at the expense of his reputation for honesty." Other newspapers offered similar assessments. The *New York Times* criticized Cook for "lecturing to credulous thousands and making a great deal of money out of a deeply clouded title." The *New York Evening Post* complained that Cook was deluding the public into "delivering both honors and dollars without due warrant." Cartoonists also lampooned Cook's profit-making campaign. In "The Arduous Return," the *St. Louis Post-Dispatch* depicted Cook trekking over a precarious terrain of white banquet tables. Later the *Post-Dispatch* offered readers an image of Cook collecting money generated from a conveyer belt powered by a wrathful Peary. The *Cleveland Leader* portrayed Cook walking by a portrait of Christopher Columbus and trailed by a porter carrying bags full of loose cash and book contracts. Responding to the label under Columbus's portrait. which reads. "Christopher Columbus / Discoverer of America / Died in Poverty." Cook remarks coolly, "It's your own fault, Chris. You didn't make the proper arrangements with the publishers." These cartoons reached beyond local audiences when they were republished in books and magazines such as the *Literary Digest*. Peary saw Cook's lecture tour in more personal terms as the theft of receipts that should have been his. Cook "has deliberately and intentionally stolen thousands of dollars," he wrote in a letter to one backer, "by as fraudulent a trick as was ever practiced by a green goods man or confidence sharp."[35]

FIGURE 16. "As the Matter Stands," *Literary Digest*, 25 September 1909, 466.

With Cook's image tarnished by these criticisms, Peary and his supporters found greater success in impugning his rival's moral character. In particular, they collected testimony from two men who had accompanied Cook to Mt. McKinley in 1906 and who claimed that he lied about his ascent. If statements by Cook's Eskimo companions failed to catalyze broad condemnation of him, these signed affidavits by two white men had greater effect. Although Peary's men had paid the two handsomely for their affidavits, a matter reported in the press, the charges took the wind out of the sails of pro-Cook newspapers such as the *New York Herald*, which soon scaled back their coverage of the controversy. In the light of growing questions about Cook's claims, his decision to continue lecturing only reinforced the view that he

FIGURE 17. "More Annexation Troubles," *Literary Digest,* 2 October 1909, 514.

was capitalizing on his discovery instead of defending his honor. While the country now looked to him to answer questions about his two expeditions, the *Literary Digest* marveled, Cook "continue[d] to reap a golden harvest on the lecture platform."[36]

SAVAGES AND CHILDREN

As Americans praised and parodied both explorers for their civilized behavior, they also evaluated each man as a frontiersman, one who had abandoned the niceties of civilization in pursuit of his goal. This is not altogether surprising considering the extensive use of primitivist rhetoric in both men's campaigns. Since the early 1900s, Peary had associated himself and his ship, the *Roosevelt*, with the rough-riding image of Theodore Roosevelt. He had played down the role of machines that made his arduous trek easier and that mediated the struggle of man against nature. Cook had taken such images one step further, representing himself as a savage on his return from the Arctic in 1909. "The reason for my success," he told a *New York Herald* reporter in 1909, "is that I returned to the primitive life–in fact, became a savage, sacrificing all comforts to the race for the pole."[37]

Published opinion about the character of each man, however, colored the way Americans interpreted their "descent" into the primitive world. In the early days of the controversy, when Peary's statements had inclined most Americans to favor Cook, many newspapers praised Cook's descent into savagery as a return to the simple virtues of an earlier, less complicated

FIGURE 18. "A Suggestion to Manufacturers of Grotesque Toys for the Holidays," *Literary Digest*, 30 October 1909, 711. The Cook-Peary controversy damaged the credibility of both explorers. Although they had once gained support for their primitive style of exploration, their public squabbling soon led reporters and satirists to portray them using traits such as childlike behavior that they associated with "primitives."

era. Echoing the language Roosevelt used to describe physical education for white men, the reporter William Stead wrote of Cook's Arctic transformation as a process in which "it seemed as though all the cells of his body and brain were burned out and replaced in the fire of that strenuous life." The reporter Philip Gibbs described Cook as "typical of Anglo-Saxon explorers, hard, simple, true." Others viewed his transformation as something more profound, something that transcended race. "He succeeded," wrote one observer in the *American Review of Reviews*, "because he became, to all intents and purposes, an Eskimo." But whether or not audiences thought Cook metamorphosed into a medieval Saxon or a stone-age Eskimo, they embraced his premodern persona as a testament to the limits of modern culture. The *Louisville Herald*, for example, viewed his voyage as a triumph of human willpower over modern machines. In language that could have been crafted by Peary himself, it wrote: "The greatest achievements of the race will be won, neither by drifting nor by flying, but by courageous determined struggle against hardship and adverse circumstances." Private citizens offered similar assessments.[38] Adding the final brush strokes to this portrait, the *New York Herald* wrote about Cook's childhood in New York as if he were already a budding frontiersman or young disciple of the Arts and Crafts movement:

> He would depend upon no one where he felt that his own efforts would result in his obtaining what he sought. So much was this so that young Cook would not ride on a "store bought" sled when his companions went out for a sport on the inclines covered with snow the present day conqueror of the Arctic went off into the woods back of his home with an ax and cut down young trees. When he had carted them into the open country where the homestead was situated he went patiently to the labor of fashioning them into a "bunker." When he had finished, he had a sled of such quality that there was none in the country to equal it.[39]

The press and the public also used primitive imagery to characterize Peary but did so in a less complimentary manner. If writers had cast Cook in the virtuous image of a noble savage, they cast Peary as his doppelgänger, a man whose exposure to savage life had brought out his uncouth, brutish nature. Even before his controversy with Cook, Peary's long years in the Arctic had raised concerns about his ability to adapt to life back home. When he lobbied Washington elites to nominate him to be the secretary of the Smithsonian Institution, a prominent member of the search committee, Andrew Dixon White, worried that Peary's long exposure to the primitive world had blunted his skills as a leader of civilized men. "White has the idea from his experience with other American explorers," Gilbert Grosvenor confided to Peary in 1907, "that the isolation in the north warps a man's disposition so that when the arctic explorer gets back to this country, he is out of touch with men and can't get on with anyone." Now the conduct Peary displayed in his controversy with Cook seemed to confirm White's suspicions. The *Camden Post-Telegram* criticized "selfish, bearish demeanor." When the men of Peary's crew refused to discuss anything related to Cook, one *New York World* reporter concluded that "Peary exercises an almost savage dominance over the members of his expedition." The *Savannah Mercury News* criticized Peary's "snarling and ungracious attitude." Peary made his statements about the controversy, reported the *World* correspondent, with a "wolfish grimness" and "an inflection that was almost a snarl."[40]

Such menacing portrayals of Peary soon gave way to more comical ones, depicting him as a petulant youth. The explorer Otto Sverdrup declared in the press that "Commander Peary [acts] like a very angry and ill-bred child." The *Denver Republican* said of Peary: "His conduct since he became aware of Dr. Cook's claim has been childish." American readers learned that even Londoners condemned his infantile behavior: "It is just like the angry howl of a baby that has had his candy stick snatched from him." The *Atlanta Journal* declared that Peary "has so completely played the baby act that nothing he could possibly say in his effort to discredit Dr. Cook will have the slightest tendency in that direction." The *Pittsburgh Independent* published a poem titled "Cook vs. Peary" that emphasized a similar theme: "So we're saying this to Peary / Better take the matter mild / Than to make the public weary / Acting like an angry child."[41]

As doubts about Cook's character grew, the press and the public began to characterize him as a child as well. Articles and cartoons started portraying the controversy as a juvenile spat. The *Literary Digest* published a cartoon of Cook and Peary as small fur-suited children complaining about each other's misdeeds to a fatherly Uncle Sam (see fig. 16, p. 149). The *South Bend Tribune* portrayed the two explorers pulling at opposite ends of the North Pole, both shouting, "It's mine." The controversy itself took on the childish qualities of its participants. The *New York Mail* portrayed the North Pole discovery as an unwanted baby left on the steps of a wary Uncle Sam. A cartoon printed in the *Literary Digest* appeared to continue the scenario, with a bewildered Uncle Sam pacing back and forth with a screaming child labeled "North Pole Controversy"[42] (see fig. 17, p. 150).

At first glance, the evolution of Cook and Peary from savage frontiersmen to crying children seems incongruous. In fact, it is quite possible that editors and satirists used such childlike imagery *because of* its incongruity with the rugged Rooseveltian images of the two explorers (see fig. 18, p. 151). However, there may have been a deeper connection between the images. By associating themselves so closely with primitive Eskimos, Cook and Peary made it easier for critics to label them as children when they appeared to reveal weakness of character. Before the Civil War, as chapter 2 makes clear, Americans already had grown accustomed to viewing native peoples as the moral and developmental equivalents of white children. If earlier writers had been interested in describing the behavior of primitive peoples using the familiar behaviors of children, now they paid more attention to their own children and the lessons to be learned from their behavior as "savages." Darwin's cousin, Francis Galton, viewed civilized children as windows into humanity's savage past. "The children who dabble and dig in the dirt," he wrote in 1865, "have inherited the instincts of untold generations of barbarian forefathers, who dug with their nails for a large fraction of their lives."[43]

Fears about overcivilization in the last decades of the nineteenth century gave new relevance to the connection between the child and the savage. It also tied it more closely to themes that Cook's and Peary's writings had put into play. In an article titled "The Savagery of Boyhood," the *Popular Science Monthly* admitted that boys' savage behaviors represented a necessary, if disagreeable, stage in their development into civilized men. "The healthy child [has], as we should expect when we consider his ancestry, the mind of a savage." Karl Pearson wrote in his biography of Francis Galton, "In youth friends are as primitive tribes[;] they raid each other's preserves both to destroy and to capture what they do not themselves possess; they mold each other's mental growth by friction and combat." Not surprisingly, the same Americans who cheered frontiersmen and explorers as role models for overcivilized urban men also encouraged boys to release the savages within. The Clark University psychologist G. Stanley Hall urged that boys give in to their normal inclination to "savage" play because it served as a developmental stage necessary for their progression into civilized adults. The founder of the Boy Scouts, Ernest Thompson Seton, organized his treks into the wilderness on the premise that boys needed to encounter the primitive world in order to become strong, civilized men. And in an article titled "The American Boy," Theodore Roosevelt championed "rough pastimes and field-sports" as solutions to the "effeminacy and luxury of young Americans" who lived in the city. As a result, the demands for children to release their inner savage closely echoed the calls of adult explorers such as Cook and Peary to "tear off the masks of civilization."[44]

In the end, Peary eked out a victory in his war of character with Cook. His success resulted more from Cook's suspicious behavior than from a restitution of his own. Amid calls to prove his claims about ascending Mt. McKinley and reaching the North Pole, Cook abruptly left the country in November 1909. Supporters claimed that he did so to recuperate from exhaustion brought on by the controversy, but newspapers roundly interpreted his departure as a way to dodge tough questions about the expeditions. Cook tried again to turn the tide of popular opinion by appealing to the Danish authorities who had so warmly received him in September. He presented Copenhagen officials with a set of records in hopes that they would verify his claim of discovery of the North Pole. The Danes concluded, however, that the records did not contain "any proof whatsoever of Dr. Cook having reached the North Pole." This repudiation of Cook's claim effectively destroyed his efforts to improve popular opinion about him back home. Meanwhile, Peary had gained some credibility from a report from a committee of the National Geographic Society that confirmed his claim of discovery of the pole. The committee reported that it reached its decision after it "carefully examined" Peary's instruments and observations. A later investigation revealed that committee members only briefly examined Peary's diaries and instruments, that some of them were longtime admirers of Peary, and that they had been predisposed to validate his claims of discovery. Yet it seemed to be enough. Combined with Cook's failure to convince the Danish commission, the committee report gained for Peary the approval of other geographical societies and eventually of Congress, which honored him for his accomplishment.[45]

The Cook-Peary controversy established Peary as the discoverer of the North Pole, but it also denied him the popular acclaim that he had so long and so anxiously sought. By raising questions about Cook's claim of discovery, he turned the spotlight on his own record and conduct as an explorer. As a result, the feud between Cook and Peary brought the idealistic images of explorers so often promoted in narratives and press reports into sharp relief against their competitive, mercenary actions at home and in the Arctic. Cook's exalted status as an American hero crumbled, leaving him a social outcast. Peary fared better, gaining the accolades of geographic societies, but these did not dispel doubts about his record, which continued to cloud his reputation and his claim to discovery of the pole. It left the explorer feeling bitter and, ironically, convinced him of the truth of the rhetoric he had so long promoted but failed to follow. "[Y]ou must make up your mind that you are doing the thing solely for the satisfaction of doing it," he wrote to a former member of his expedition, "because if you go and win and come back you will get nothing for it from the people of this country at least."[46]

Some resisted this ignoble end to U.S. discovery expeditions in the Arctic. Fans tried to burnish the image of explorers by comparing their debates to other momentous feuds in history. One woman wrote a poem to Cook encouraging him to see the controversy as an inevitable consequence of his great discovery. "'Tis always so with greatness / And pity that 'tis so / Columbus wore the iron of chains / And kiss'd them in his wo." Writers for the popular press also used this approach. In his article "The Disputes of Great Discoverers," Harry Thurston Peck wrote, "The explorer and discoverer may expect to have his word discredited. . . . This is the sacrifice that is often exacted from those who seek the truth. They must look to distant ages for a late reward." Edwin Swift Balch, author of one of the first books about the expeditions to the North Pole, also defended the controversy as a common manifestation of great discoveries: "From time immemorial, travelers have been called liars. The number of those who have been told that they were fakirs and had handed a gold brick to the public, or the equivalent of such a statement, and whose discoveries nevertheless have been verified in due time, is legion."[47]

But to most reporters of the day, the story of Cook and Peary offered a different moral, simple and compelling. It warned of dangers for men who pursued popular glory without regard to their personal scruples. In this sense, the controversy appeared to be a drama governed by the temperaments of two men rather than one shaped by the press, the attitudes of the public, or ideas about modern exploration. So although the public came to see explorers as motivated by more than a desire to go it alone in the Arctic, they did not see the controversy's deeper roots. It remained an ugly affair that reflected badly on explorers but seemed to have little to do with American society and its values.

CONCLUSION

Analysis of the controversy makes clear, however, that though Cook's and Peary's actions may have sparked the disagreement, it was fueled by conflicting cultural and commercial forces. In particular, the explosive rise of popular media and advertising in the 1890s provided explorers with new means of promotion and new sources of revenue. Never before had lectures and narratives proved so lucrative for them. Yet never had commercial opportunities posed such grave risks. The rise of advertising and mass consumerism in the 1890s raised fears of moral decline among many white Americans and generated interest in rustic, nostalgic heroes who had turned their backs on the emasculating forces of modern society. Cook and Peary exploited these opportunities to great effect by representing themselves as rugged, anticommercial figures. Their disavowal of science and modern inventions and their close association with the Eskimos only enhanced their ruggedness while distancing them from the commercialism of American culture. Before the controversy of 1909 began, then, the stage was set for a clash between explorers' actions and their images as public figures. When Peary lit the flame of controversy with his remarks, the popular press eagerly pursued the story as zealously as it had covered Greely's expedition twenty-five years earlier. It had learned well from James Gordon Bennett that scandalous expedition stories sell just as well as, if not better, than heroic ones.

If the controversy that engulfed Cook and Peary left the discoverer of the North Pole in doubt, it proved more revealing of the conflicting expectations to which twentieth-century explorers had to conform. On one hand, their audiences continued to see them as representatives of national progress, and their discoveries added to the glory of the nation. On the other, a number of Americans began to hold up Arctic explorers as premodern figures, antidotes to progress rather than manifestations of it. As the controversy gained steam, both explorers drew criticism for their traits as men of civilization and as men of the wild frontier. Diverse writers and intellectuals believed that primitive and civilized traits could be, indeed, must be, blended within the character of the modern man. In the early days of the controversy, when Cook had gained the greatest sympathy of his American audiences, he seemed to embody the ideal mixture of modern and premodern traits. As audiences grew disenchanted with him, however, such blending of traits came to appear inconsistent and even hypocritical. In a remark that became an epigram for the Cook-Peary controversy, New York senator Chauncey Depew told the Transportation Club that "Cook is a liar and a gentleman, and Peary is neither." The incongruity of the phrase seemed to capture the strange debate and was widely reprinted for years afterwards. It also captured the conflict in the ideals to which both explorers had been held. In the years after the controversy, both men considered new expeditions to resuscitate their reputations as explorers. But soon thereafter, the outbreak of war in Europe provided Americans with new ranks of manly heroes, martyrs, and patriots with which explorers could not compete.[48]

CONCLUSION

——— ✳ ———

"EXPLORERS are like prisoners," Frederick Cook wrote in his memoir sometime in the 1920s. The Arctic's long winters and brutal cold held them captive, frayed their nerves, and sent them into a "cloud of incipient insanity." In his view, the Arctic did not build character, as explorers had long declared, but corroded it, sapping men's strength and spirit "out of the darkness of denied freedom." The metaphor of the prisoner was at odds with Cook's earlier writings, which had depicted the Arctic as a place of liberation, a haven from the confinement of civilized life. But it was particularly fitting for his life in the 1920s. After World War I, Cook had used his Arctic celebrity to sell shares in dubious oil businesses and was convicted of mail fraud in 1923. As he wrote, he was a prisoner himself, incarcerated at Leavenworth Penitentiary.[1]

The metaphor of the prisoner proved fitting in another sense. Explorers had become shackled to an impossible set of expectations as public figures. By 1900, they had cut themselves off from scientists and scholars. They justified their expeditions using the slimmest of pretexts: tests of manly character. They made little of commerce, science, or other pursuits. As character became all-important, reports of explorers' bad behavior proved especially damaging. But the decline of Arctic explorers as public figures cannot be chalked up entirely to their poor conduct. In the 1850s Elisha Kane's lapses of leadership were well known to his officers and crew, but they did little to prevent his transformation into a hero and martyr. By contrast, after 1880, *most* Arctic explorers came under fire for questionable actions. The pace of scandal quickened because of the press's willingness to investigate the dark side of expeditionary life. But scandals also fed on the uncertainty about manly character itself. Its qualities were ephemeral, changing to keep pace with American culture. If explorers failed to live up to standards of character, it was, in part, because the standards themselves were in flux.

EXPLORERS AND SCIENTISTS

Scientists watched explorers fall off the pedestal of public opinion with some relief. "The way is now clear," stated the science journal *Nature*, "for the scientific study of arctic hydrography, meteorology, and many other problems of terrestrial physics without the disturbing effort to attain the highest latitude." Even scientists who had once worked closely with explorers welcomed the end of the quest for the North Pole. "We must not forget that the explorer is not expected merely to travel from one point to another," wrote the anthropologist Franz Boas, "but that we must expect him also to see and to observe things worth seeing." Much had changed since the 1850s, when scientists had been explorers' staunchest allies. In those days, Louis Agassiz, one of America's most famous men of science, had defended Arctic exploration as "highly important . . . from a scientific point of view."[2]

The story of this estrangement is important because it tells us about Arctic exploration as a cultural phenomenon in America. During the early days of U.S. polar exploration, scientists involved themselves in the nuts and bolts of expeditionary campaigns. They obtained scientific instruments, raised money, and lobbied members of Congress. More important—from our perspective—they functioned as a cultural barometer of Arctic exploration, a gauge of explorers' campaigns and their popular appeal. Understanding the cultural role of science, then, gives us a better fix on the position of explorers in American life.

Yet Arctic exploration is also a yardstick by which we can measure the cultural position of American science. When we do so, new features appear. The early decades of Arctic exploration held few surprises. Elisha Kane thrilled popular and scientific audiences alike, a feat that confirms the work of American historians who have written about the close links between amateurs and elites in the 1850s. More surprising is the falling-out between explorers and scientists in the 1880s and 1890s, a time when the reputation of scientists had never seemed greater. Why would explorers turn away from scientists just at the moment when scientists were coming of age as cultural authorities?

There are many reasons. The rise of the popular press after the Civil War provided new opportunities to fund Arctic exploration. Explorers often found it easier to attach themselves to wealthy commercial patrons than to woo the scientific community. For their part, scientists grew increasingly distrustful of explorers' claims. The rapid professionalization of American science in the late nineteenth century widened the cultural chasm between scientists and explorers and made a rapprochement difficult. In the field of geography in particular, scholars had grown weary of adventurers who passed off their work s as serious research. "The explorer so often directs his attention to other subjects than careful description of the lands over which he travels," wrote the Harvard University geographer William Davis, "that he cannot in all cases be called a geographer." Although amateur explorers continued to find a home at the National Geographic Society (which now attended full time to the public's interest in geography), they had few friends in professional organizations such as the Association of American Geographers. Whereas Kane had mingled easily with elite scholars in the 1850s, turn-of-the-century explorers rarely crossed their paths, wrote for their journals, or attended their meetings.[3]

Yet these are only partial explanations. By 1890, explorers had found new niches as cultural figures that made it difficult for them to identify with scientists. In the 1850s and 1860s, American audiences held up the first generation of Arctic explorers as men of the future, individuals who revealed American society at its most progressive. Science fit well with the image of explorers as forward-looking figures; they impressed audiences with their worldly knowledge, eloquence, and cultivation. Fifty years later, however, they had become figures out of a glorious past, captivating audiences because of their rejection of modern civilization and its excesses. For explorers playing these new roles, science seemed increasingly out of character. Discussions of Arctic research did not complement their image as rugged frontiersmen and primitive savages. However, the image of the Arctic explorer did not change smoothly or consistently. As the campaigns of Wellman and Peary make clear, the 1890s and early 1900s were a time in which explorers operated as progressive and nostalgic figures. In the case of the North Pole controversy of 1909, both Cook and Peary sometimes exhibited modern and old-fashioned attributes at the same time. For their part, some men of science also began to exhibit a mixture of modern and muscular traits, gaining status among their peers not only for their observations but also for their arduous efforts to obtain them.[4]

Despite this mingling of manly attributes, the ideals guiding explorers grew apart from those used by scientists. In particular, turn-of-the-century scientists were figures who were still praised for their rationality and disinterestedness, traits that also identified them as men. While critics of progress accepted these as male traits, they viewed them as repressive ones: behaviors that men adopted to keep their animal passions at bay. Men who sought to liberate themselves from the emasculating effects of civilization, therefore, had to find new ways to behave, or, at the very least, new ways of being rational. For disciples of the new masculine ethos, Anthony Rotundo observes, "male rationality was not a capacity for deep, logical reflection but rather an absence of complex emotions—an absence which freed men to act boldly and decisively." Peary certainly cultivated this new form of rationality. Whereas Kane's writings overflowed with insightful analysis and colorful descriptions, Peary's writings were lean and unemotional. Only violence, not visions of nature, seemed to rouse the narrator from his emotional slumber. "Twenty-odd years of arctic experience," he admitted in *The North Pole*, "had dulled for me the excitement of everything but the polar bear chase."[5]

ARCTIC FEVER

If Peary had become dulled to the Arctic itself, how did he justify his obsessive pursuit of Arctic exploration? As he tells it, he could not help himself. "The northern bacilli were in my system, the arctic fever in my veins, never to be eradicated." Explorers increasingly blamed "Arctic fever" for the strange compulsion to go north against all common sense. Emma De Long wrote of her husband George that "the polar virus was in his blood and would not let him rest." All first-time Arctic explorers, Cook wrote, "were infected with the same spirit." Even audiences were not immune to it. "Arctic enthusiasm is an intermittent fever," stated *McClure's Magazine* in 1893, "returning in almost epidemic form after intervals of normal indifference." By 1900, outbreaks of Arctic fever became increasingly common, and explorers self-diagnosed it with great regularity.[6]

The disease may seem to be nothing but a playful literary metaphor, but it had serious functions. Arctic fever located the urge to explore in the human passions. It was a condition that afflicted the heart against the better judgment of the mind, operating beyond conscious control. Why should anyone attempt to reach the North Pole when it served no useful or scientific function? They did so because, explorers claimed, they felt irrationally compelled. In this way, Arctic fever masked rational motives for voyaging north, namely, the promise of celebrity and financial reward. This explains why Arctic fever broke out so late in the nineteenth century. By 1900, practical reasons for exploring the Arctic (such as the search for the Northwest Passage and the rescue of John Franklin) had all but disappeared. The idea of feverish compulsions fit in well with the new image of explorers who acted on their instincts rather than on reasoned calculation. In 1857, eulogists had praised Elisha Kane (the only explorer to suffer from real fevers) for his great passion. Yet Kane's passion was of different stuff than Peary's. It was a Romantic sensibility that led Kane to write about the polar regions with the self-conscious wonder of *Walden*. By contrast, Cook and Peary expressed passions that were torn from the pages of *Call of the Wild*, as if Arctic exploration were an impulse rooted deep within the primate brain.

Despite the scandals that engulfed explorers in the late nineteenth century, Arctic fever proved remarkably difficult to contain. Only during the North Pole controversy of 1909 did Americans categorically question the motives that drove explorers into the polar regions. Although they grew cynical regarding Cook and Peary, they did not entirely lose faith in the project of exploration. By the end of World War I, many had reawakened to the lure of the polar regions. Americans cheered Richard Byrd's flight over the North Pole[7] in 1926 and his flight over the South Pole in 1929. The success of such expeditions in the face of earlier scandals suggests that audiences had not lost faith in the basic tenet of Arctic fever: that a pure—perhaps irrational—impulse drove explorers to roam the dangerous places of the world. In fact, Arctic fever still rages today, despite the recent avalanche of works detailing explorers' behind-the-scenes activities and and real-world motivations. It persists even within scholarly studies of exploration. Writing about expeditions to the North Pole in *A Continent Comprehended*, one of the most comprehensive works on North American exploration, W. Gillies Ross observes: "That so much effort should be expended, and so much hardship endured, in a quest for this featureless point, set within a frozen ocean, is vivid testimony to the irrational element in exploration."[8]

More than any event or institution, this ethos has been Arctic explorers' most lasting legacy today: the widespread belief that exploration is an impulsive and instinctive activity, deeply rooted within the human psyche. The idea has a complicated genealogy, to be sure. Arctic fever has cultural antecedents that date back to the writings of Renaissance pilgrims and Romantic travelers. Moreover, it was not confined to the Arctic but existed as part of a larger pandemic that afflicted Western explorers in Africa, South America, and the South Pacific. Their narratives, too, admit to a feverish impulse to explore.⁹ Yet Arctic exploration did not merely reflect ideas swirling through American—or Western—culture. It was an activity that shaped the attitudes of middle-class Americans. After the rapid settlement of the western territories, Americans invested the Arctic with a special status as a frontier. It was in Arctic campaigns that the idea of the obsessive explorer gained its fullest expression. And it was in debates about polar voyages that scientists and explorers aired their strongest grievances with each other and articulated different visions of exploration. If Arctic fever reflected ideas and attitudes already present in American culture, it also projected them, focused them, and gave them the force of argument.

So it is not surprising that we should see new incarnations of Arctic fever, adapted for other regions and geographical projects. Only a decade after the end of the race for the poles, George Mallory came down with the mountaineer's strain of the illness, professing an urge to scale the world's highest peak "because it is there." For the historian J. R. L. Anderson, Arctic fever was simply a particularly virulent form of "the Ulysses Factor," a condition characterized by "deliberate risk-taking in pursuit of a goal of no apparent practical value."¹⁰

Nowhere is the legacy of Arctic Fever more visible than in the U.S. space program. Although NASA administrators are at pains to show that human space flight is both practical and scientific, they also defend it as an irresistible impulse rooted in the human condition. "The cause of exploration is not an option we choose," reads the epigraph to its 2004 report *The Vision for Space Exploration*. "It is a desire written within the human heart." The statement is taken from a speech by President George W. Bush at a memorial service for the astronauts aboard the space shuttle *Columbia*. But it could have been written by Robert Peary himself. Dressed as an observation about the human condition, Bush's statement conceals a specific argument for a particular policy: humans need to explore the solar system (in this case, Mars) even when robots could probably do the job more cheaply, safely, and effectively.¹¹ Across the distance of time and culture, Peary would have understood this argument at once: machines cannot replace human explorers—not because they cannot do a good job but because they cannot, ultimately, sustain the attention of a human audience. And so, some things change enough to come full circle. Explorers today keep a close eye on home audiences even as they hurtle into space. Freed from the bonds of gravity, they still remain tied to the needs of a watchful public.

Notes

1. For many native peoples of the Arctic today, *Eskimo* is an offensive term. I use it here and elsewhere because it is the term commonly used by nineteenth-century Americans.

2. Peary was speaking to Mary French, who quoted him in her manuscript "Memories of a Sculptor's Wife," 2, box "Diary Transcripts, 1871–1900," Robert E. Peary Papers, National Archives, College Park, Maryland (hereafter Peary Papers). Details of the expedition come from Peary, *North Pole*, 9, 74, 119 (quotation), 213.

3. Details of Peary's activities in the months before his departure come from "Letters Received," 1907, Peary Papers. Wally Herbert discusses Peary's post-expedition plans in *Noose of Laurels*, 239–240.

4. Although I focus mostly on events at home, I owe a debt to expedition historians who, in the past thirty years, have scrutinized Arctic explorers' accounts and placed them in broader context. Their works correct many of the exaggerations and omissions of earlier accounts and inform many of my chapters. See Corner, *Dr. Kane of the Arctic Seas*; Loomis, *Weird and Tragic Shores*; Guttridge, *Icebound*; Guttridge, *Ghosts of Cape Sabine*; Bryce, *Cook and Peary*.

5. Rubin, *Making of Middlebrow Culture*, 2–10. On antebellum ideals of character, see also Manthorne, "Legible Landscapes"; Bell, "Conditions of Literary Vocation," 74; Miller, *Empire of the Eye*, 35; Morse, *American Romanticism*, 28.

6. Henry, *Scientist in American Life*, 37, addressing the American Association for the Advancement of Science in 1850. On the relation between science and character in antebellum America, see Keeney, *Botanizers*, 38–41; Slotten, *Patronage, Practice, and the Culture of American Science*, 24–27.

7. Among historians, this more expansive notion of science has many advocates. For a good discussion of the issue, see Cooter and Pumfrey, "Separate Spheres and Public Places." Increasingly, scholars such as Cooter and Pumfrey have urged researchers to turn their attention away from scientific elites to focus on "the ordinary men and women who form the stuff of social history." In the case of Arctic exploration, such an approach suggests a way of overcoming a rift in the scholarship regarding the role of science in polar voyages. On one hand, historians of science and geography have concentrated on Arctic science in the field, arguing that the popular treatments of such expeditions have overshadowed the importance of their field work. On the other, scholars

who study exploration literature have tended to discount Arctic science as the empty rhetoric of explorers' campaigns. In my opinion, both views emerge from a restricted view of science as a field practice, exclusive of its ability to function as a social activity or system of meanings. Accepting rhetoric as an important component of scientific work helps us make sense not only of changing ideas about the Arctic but also of the changing status of the explorers who traveled there.

8. On science as a male preserve, see Rossiter, *Women Scientists in America*. For scientists' use of late nineteenth-century manly ideals see Herzig, "In the Name of Truth"; Hevly, "Heroic Science of Glacier Motion."

9. "Our Friend the Eskimo," *New York Sun*, 15 September 1912, 10. Explorers did retain some support after 1911, particularly among diehard geographical societies.

10. R. Peary, "Remarks to the National Geographic Society," 1907, Peary Papers.

11. J. Peary, *Snow Baby*.

12. Browne, *Charles Darwin*; Driver, *Geography Militant*, 8; Rozwadowski, *Fathoming the Ocean*.

CHAPTER ONE

1. On early North American surveys, see Meining, *Shaping of America*; Hindle, *Pursuit of Science*; Goetzmann, *Exploration and Empire*; Daniels, *Science in American Society*. On Thomas Hutchins, see *American National Biography*, s.v. "Hutchins, Thomas." The quotation is from *Journal of the Continental Congress* 29 (18 May 1785): 923.

2. Thomas Jefferson to James Monroe, 26 May 1801, Thomas Jefferson Papers, Manuscript Division, Library of Congress, Washington, DC (hereafter Jefferson Papers).

3. Thomas Jefferson to Congress, 18 January 1803, Jefferson Papers.

4. See La Pérouse, *Journal of Jean-François de Galaup de la Pérouse*.

5. Thomas Jefferson to Congress, 18 January 1803.

6. Spencer, "'We are not entirely dealing with the past.'" On the delay in publication of expedition documents, see Dupree, *Science*, 26-28.

7. Goetzmann, *Army Exploration*, 34-37; Goetzmann, *Exploration and Empire*, x-xiii, 3-8, 181-198; Dupree, *Science*, 56.

8. Slotten, *Patronage*, 68-82; Bruce, *Launching of American Science*, 171-189; Dupree, *Science*, 32, 55-56, 101-102.

9. Viola and Margolis, *Magnificent Voyagers*, introduction; Dupree, *Science*, 56.

10. The Harper executive was speaking to John L. Stephens, who recounted the conversation in Thomas Law Nichols's *Forty Years of American Life*, 343. I first learned of the exchange in Larzen Ziff's *Return Passages*, 8. Examples of early exploration literature include Irving, *Adventures of Captain Bonneville*; Poe, *Narrative of Arthur Gordon Pym*; *Mariner's Chronicles*. For discussion of these works, see Lenz, "Narratives of Exploration."

11. Irving, *Astoria*; Cooper, *Pioneers*. For examples of Cole's work, see *Mountain Sunrise* (1826) and *The Oxbow* (1836). For examples of Church landscapes further from home see *The Andes of Equador* (1855) and *Morning in the Tropics* (1877).

12. The quotation is from T. Cole, "Essay on American Scenery." On the changing ethos of American landscape painting, see Madsen, *American Exceptionalism*, 70-99; Novak, *Nature and Culture*, 139-151; Miller, *Empire of the Eye*; Manthorne, "Legible Landscapes," 133-145; Manthorne, *Tropical Renaissance*, 133-156. For examples of early American themes in art, see the Revolutionary War paintings of John Trumbull (1756-1843) and the portraits of Charles Willson Peale (1741-1827) and Gilbert Stuart (1755-1828).

13. After the voyages of Baffin in 1616, the search for the Northwest Passage slowed but did not come to a halt. James Knight, Christopher Middleton, Samuel Hearne, James Cook, and Alexander

MacKenzie also sought the passage in the eighteenth century. Savours, *Search for the North West Passage*, 23–38. Barrow is quoted in Fleming, *Barrow's Boys*, 11.

14. Shelley, *Frankenstein*, 248; Bragg, *Voyage to the North Pole*, 4. On early British expeditions, see Riffenburgh, *Myth of the Explorer*; Holland, *Arctic Exploration and Development*; Mirsky, *To the Arctic*, 97.

15. Tebbel, *History of Book Publishing*, 208; Hawkes, *Uncle Philip's Conversation*. Examples of reprinted narratives include Parry, *Three Voyages*; Ross, *Narrative of a Second Voyage*; *Arctic Travels*; Back, *Narrative of the Arctic Land Expedition*; Williams, *Narrative of the Recent Voyage of Captain Ross*.

16. Sigourney, "King of Icebergs"; Bryant, "Arctic Lover to His Mistress." On Symmes, see Lenz, "Narratives of Exploration," 47–55; P. Clark, "Symmes Theory of the Earth"; Godwin, *Arktos*; Madden, "Symmes and His Theory." Wilkes included his address to Congress in the *Narrative of the U. S. Exploring Expedition*.

17. Seaborne, *Symzonia*; Poe, *Narrative of Arthur Gordon Pym*. Other examples include Bird, "This Ice-Island"; Cooper, *Sea Lions*. Poe is quoted in Lenz, "Narratives of Exploration," 56.

18. The literature concerning the Franklin search is vast. For some good examples, see Levere, *Science and the Canadian Arctic*; Caswell, *Arctic Frontiers*; Ross, "Nineteenth-Century Exploration of the Arctic"; Neatby, *Search for Franklin*. British and American mediums frequently reported news and sightings of Franklin. On the vision of Franklin in Japan, see "Sir John Franklin," *Boston Daily Evening Transcript*, 18 May 1850, 1. On the broader issue of Franklin and the supernatural, see Ross, "Clairvoyants and Mediums."

19. Corner, *Dr. Kane*, 76–77. Nineteenth-century indices of American newspapers and periodicals confirm a dramatic increase in press coverage during the late 1840s. For examples, see indices of *Harper's New Monthly Magazine* and *Littel's Living Age*.

20. Baker, "Franklin Expedition"; Butler, "Sir John Franklin Expedition."

21. Bayly, "Franklin Expedition," 833; Savage, "Franklin Expedition," 832; "Sir John Franklin," *Philadelphia Public Ledger*, 2 May 1850, 2; Seward, "Sir John Franklin Expedition," 885–886.

22. David Chapin discusses the political windfalls of Arctic exploration in "'Science Weeps.'" The quotation is from John Cable's resolution, reprinted in "Franklin Expedition." The Senate confirmed the resolution by a vote of 28 to 16. The House approved it by a vote of 94 to 45.

CHAPTER TWO

1. Elder, *Biography*, 306–307; "Remains of Dr. Kane," 3; Joint Committee, *Honors to Dr. Kane*; Elder, *Biography*, 308, 345 (quotation). Other historians have also read larger meanings into the pageantry of Kane's funeral. See ibid.; Joint Committee, *Honors to Dr. Kane*; Mirsky, *Elisha Kent Kane*, 3–12; Chapin, "'Science Weeps.'"

2. See eulogies in Joint Committee, *Honors to Dr. Kane*; Alger, *Brief Tribute*; Andrews, *Memoir and Eulogy*; Chamberlain, *Dr. Kane and Christian Heroism*; Shields, *Funeral Eulogy*.

3. On Kane's cortège see Elder, *Biography*, 306–308, 345; "Remains of Dr. Kane," 3; Joint Committee, *Honors to Dr. Kane*. Quotations of eulogists are taken from Elder, *Biography*, 324 (Charles Anderson), 398 (C. Edwards Lester) 398; Joint Committee, *Honors to Dr. Kane*, 26 (James Batthe).

4. Corner, *Dr. Kane of the Arctic Seas*, 18–70; Elisha Kane to Jane Leiper Kane, 16 January 1850, Elisha Kent Kane Papers, American Philosophical Society Library, Philadelphia (hereafter Kane Papers).

5. Kane, *United States Grinnell Expedition*, 151–160; "Sir John Franklin and the Arctic Regions," 180; "The Grinnell Franklin Exploring Expedition," *New York Herald*, 22 May 1850, 4; "Sailing of the Grinnell Arctic Expedition," *New York Herald*, 24 May 1850, 4; "Mr. Grinnell's Arctic Expedition," *Boston Daily Evening Transcript*, 23 May 1850, 2.

6. "The Search for John Franklin," *New-York Daily Times*, 14 February 1852, 4.

7. Wright, "Open Polar Sea"; Godwin, *Arktos;* De Haven, "Official Report," 498.

8. Kane, "Sir John Franklin: His Probable Course"; "American Arctic Expedition," 17 (for the *Harper's* quotation). Kane's conclusions found support in the reports of British Arctic commanders who also suggested that Franklin had traveled north into the open polar sea: Untitled article, *New-York Daily Times,* 26 May 1852, 2; "The Search for Sir John Franklin," *New-York Daily Times,* 25 November 1852; "Second American Expedition in Search of Sir John Franklin"; Parry, "Sir Edward Parry's Opinion."

9. D. Scott, "Popular Lecture." Records of cities requesting lectures by Kane are scattered throughout the Kane Papers and include Wilmington, DE, Boston, MA, Plymouth, MA, North-bridge, MA, New Bedford, MA, Salem, MA, Medford, MA, New London, CT, New York, NY, Buffalo, NY, Poughkeepsie, NY, Brattleboro, VT, Baltimore, MD, Cincinnati, OH, Richmond, VA, Portland, ME, Providence, RI, and Montreal, Canada. The origin of some requests cannot be identified. H. T. Wells to Elisha Kent Kane, 18 February 1852, Kane Papers; Henry Grinnell to Elisha Kent Kane, 11 January 1852, ibid.

10. Kohlstedt, "Creating a Forum for Science: AAAS in the Nineteenth Century"; Wright, *Geography in the Making;* Bruce, *Launching of an American Science,* 172-189.

11. Slotten, *Patronage;* Maury, *Physical Geography of the Sea.* Kane credited Maury with finding biogeographical evidence that supported the theory of an open polar sea. Maury claimed that merchant sailors had sighted whales along the coast of Greenland that carried the harpoons of Pacific whalers. Since these whales had never been seen migrating southward around the American continents, the only possible path of migration was across the Arctic Ocean. Because whales had to surface for air, Maury concluded, the polar sea must be at least partially open. "New-York Geographical Society: Access to an Open Polar Sea Along a North American Meridian; A Paper by Dr. E. K. Kane, Lieut. U.S.N.," *New-York Daily Times,* 15 December 1852, 4.

12. Henry Grinnell to Elisha Kent Kane, 2 January 1853, Kane Papers (emphasis in original).

13. Allen, "Elisha Kent Kane," n.p.

14. Emma Lou Sprigman to Jane Duval Leiper Kane, 20 April 1858, Kane Papers; Shields, *Funeral Eulogy* ("transporting"), 9; Edselas, "Boston Half a Century Ago" ("brave explorer") 735.

15. Elisha Kent Kane to Charles Schott, 8 March 1852, Kane Papers.

16. *Proceedings of the Academy of Natural Sciences of Philadelphia* (1852), 298; *Proceedings of the Academy of Natural Sciences of Philadelphia* (1853), xvii.

17. Corner, *Dr. Kane of the Arctic Seas,* 120. See also Henry Grinnell to Elisha Kent Kane, 2 January 1853; Henry Grinnell to Elisha Kent Kane, 18 January 1853, both in Kane Papers.

18. Corner, *Dr. Kane of the Arctic Seas,* 154-155.

19. The final "withdrawal party" was made up of eight men. Corner, *Dr. Kane of the Arctic Seas,* 179. For Eskimo testimony regarding the same events see Harper, *Give Me My Father's Body,* 18.

20. T. L. Kane to E. K. Kane, 21 May 1855, Kane Papers (emphasis in original). "Reception of the News—Excitement in Town," *New-York Daily Times,* 12 October 1855, 8.

21. On Kane's reception, see "Reception of the News"; Corner, *Dr. Kane of the Arctic Seas,* 224-225; D. Scott, "Popular Lecture," 802. On Wallack, see *American National Biography,* s.v. "Wallack, James William." *Frank Leslie's Illustrated Newspaper* is quoted by Mott, *History of American Magazines,* 2:204.

22. Corner, *Dr. Kane of the Arctic Seas,* 235.

23. Kane, *Arctic Explorations,* 1:76 ("I could not help"); private journal, Kane Papers; Kane, *Arctic Explorations,* 1:352 ("confident"), 1:439 ("brother's welcome").

24. Kane, *Arctic Explorations,* 1:386; Joyce, "Elisha Kent Kane," 103-117; Joyce, "As the Wolf from the Dog," 333-334. Even the success of Kane's crew in hunting game was owed to native Greenlanders, Hendrik and Petersen.

25. Describing New World natives as children dates to the time of Christopher Columbus.

26. Rogin, *Fathers and Children,* 114 n. 3; Berkhofer, "White Conceptions of Indians"; Huhndorf, "Going Native"; Kane, *Arctic Explorations,* 2:118, 2:123, 2:118, 1:384.

27. Isaac I. Hayes to Elisha Kent Kane, 28 January [1856], Kane Papers; Kane's delegation of duties aboard the ship made Wilson's position ambiguous, but he came aboard with a higher ranking than that of Henry Brooks, who served as first mate.

28. Corner, *Dr. Kane of the Arctic Seas*, 236–238.

29. Shields, *Funeral Eulogy*, 8. C. Edward Lester's letter to the Masons is quoted in Andrews, *Memoir and Eulogy*, 17.

30. Elder, *Biography*, 27, 35; John Fries Frazer is quoted in Joint Committee, *Honors to Dr. Kane*, 20. For more on Frazer, see *American National Biography*, s.v. "Frazer, John Fries." Also see the remarks of Thomas H. Weasner in Elder, *Biography* (320) and those of Charles Anderson (324).

31. Elder, *Biography*, 288 (for Cuyler, the Philadelphia man), 273 (Bonsall, the officer), 283 (Goodfellow, the crewman); Morton, *Dr. Kane's Arctic Voyage*, 2. Kane regained some weight posthumously. In an 1875 history of the United States, he had fattened up to 106 pounds; see *Centennial Edition of the History of the United States*. The maid, Belle Burns, is quoted in Corner, *Dr. Kane of the Arctic Seas*, 27; *Daily Telegraph* (Harrisburg, PA), n.d., is quoted in Chapin, "'Science Weeps.'"

32. Elder, *Biography*, 271 (for Hayes), 2 (for Morton); Allen, *Elisha Kent Kane*, 2.

33. The eulogists are quoted in Elder, *Biography*, 317–318 (for the first quotation), 316 (for the second quotation). Elder, by contrast, did discuss Kane's military record.

34. "The Last of the Arctic Explorers," *New-York Daily Times*, 13 October 1855, 4.

35. Sargent, *Arctic Adventure by Land and Sea*, v; Elder, *Biography*, 292.

36. Andrews, *Memoir and Eulogy*, 266; Elder, *Biography*, 280.

37. There are exceptions. Charles Wilkes, commander of the Exploring Expedition in 1838, and John Charles Fremont, explorer of the western frontier in the 1840s, both found a measure of fame (and notoriety) in the wake of their expeditions.

38. On the Kane panoramas, see Potter, "Sublime Yet Awful Grandeur," 194. Ralph Waldo Emerson and Henry David Thoreau both mention Kane by name in their writings. For examples, see Emerson's essay "Wealth" in *Conduct of Life*, vol. 6 of his *Complete Works*, and Thoreau's 1854 essay "What Shall It Profit," later published as *Life Without Principle*. Scenes of Arctic life are present in Walt Whitman's *Leaves of Grass*, 38, and in many of Emily Dickinson's poems. See *Poems of Emily Dickinson*, 532, 792, 851, 1577. Thirty years after Kane's return from the Arctic, the Franklin search remained vivid to Dickinson, who still alluded to it in her letters. See *Letters of Emily Dickinson*, vol. 3.

39. Smith, *My Experience*; *Dictionary of American Biography* (1932), s.v. "Kane, Elisha Kent."

CHAPTER THREE

1. Rae, *John Rae's Arctic Correspondence*, 16:lxxix; Rae, "Remains of Sir John Franklin."

2. The *London Examiner* story, "The Search for Sir John Franklin," was reprinted in *Littell's Living Age*, 9 February 1850. Dickens, "The Lost Arctic Voyagers," 207. The Dickens-Rae debate unfolded in the pages of *Household Words* (December 2, 9, 23, 30), the *Athenaeum*, and the *London Times* (October 30, 31 and November 1, 3, 7). For more about the controversy, see *John Rae's Arctic Correspondence*, lxxxviii.

3. In the United States, *Littell's Living Age* followed the debate closely. See "Miscellaneous"; Dickens, "The Lost Arctic Voyagers"; Rae, "Remains of Sir John Franklin"; Rae, "Miscellaneous"; Rae, "Lost Arctic Voyagers."

4. Historians usually frame Charles Hall as a capable, if iconoclastic, explorer. The best study of Hall is Chauncey Loomis's 1971 biography *Weird and Tragic Shores*. Earlier biographies prove less critical of Hall but, for that reason, more useful as a source of exploration rhetoric. See Davis, *Narrative of the North Polar Expedition*; Nourse, *Narrative*. For more recent work on Hall, see Ross, "Nineteenth-Century Exploration of the Arctic"; Riffenburgh, *Myth of the Explorer*. Despite the recent renaissance in the history of Arctic exploration, Hayes's obscurity seems to be increasing. Although he still warranted an entry in the 1932 *Dictionary of American Biography*, he failed to

make the cut for the recent and more extensive *American National Biography* (1999). When Hayes does appear, it is in general histories of Arctic exploration, works that discuss the theory of the open polar sea, or as a secondary figure in histories of other expeditions. See Berton, *Arctic Grail;* Holland, *Farthest North;* Holland, *Arctic Exploration and Development;* Bruce, *Launching of American Science;* Carr, *Frederick Edwin Church;* Caswell, *Arctic Frontiers;* Herbert, *Noose of Laurels;* Harper, *Give Me My Father's Body.*

5. Hayes had good reason to be concerned about his portrayal as a deserter. References to the Kane "mutiny" were not uncommon in popular texts. See J. Allen, *Elisha Kent Kane.*

6. Hayes published Bache's and Henry's letters of support in Hayes, *Polar Exploring Expedition,* 6-7.

7. Hayes, *Open Polar Sea,* 4-5.

8. Hayes, "Observations." On the prominence of the *American Journal of Arts and Sciences* see Keeney, *Botanizers,* 29. For the declarations of support, see *Proceedings of the Academy of Natural Sciences of Philadelphia* (1858): 113. The report of Egbert Viele, the AGS committee chairman, is found in Hayes, *Polar Exploring Expedition,* 21. Viele's letters and others were subsequently published in the *New-York Times,* 23 March 1860, 8.

9. Hayes also persuaded Agassiz that the polar sea was open. See Agassiz to LeConte, 6 December 1860, LeConte Papers, American Philosophical Society, Philadelphia (hereafter APS). For information about subscribers to the Hayes expedition, see Hayes, *Open Polar Sea,* xi–xvi; Isaac Hayes to W. Parker Foulke, 25 April 1860, William Parker Foulke Papers, APS. Hayes reports on his success in raising funds in *Proceedings of the Academy of Natural Sciences of Philadelphia* (1861), 149–150.

10. Hayes, *Open Polar Sea,* 5-7, viii–ix; "Arctic Boat Journey."

11. Hayes, *Polar Exploring Expedition,* 11, 20; "The New Polar Expedition," *New-York Times,* 8 March 1860, 4; "The Polar Exploring Expedition," *New-York Times,* 11 April 1860, 1.

12. Quoted in Hayes, *Polar Exploring Expedition,* 27, 30.

13. "Farewell to the Arctic Expedition," *Boston Daily Evening Transcript,* 5 July 1860, 3; Hayes, *Open Polar Sea,* 12-15.

14. On the death of ship's carpenter Gibson Caruthers, see Hayes, *Open Polar Sea,* 37-38. The story of Longshaw remains murky. His name is absent from crew lists published prior to Hayes's departure. Newspapers identified him as a member of the party only after *United States* left Boston. Because Hayes expunged all references to him from his narrative, the only surviving account of this scandal comes from Henry Grinnell's letters to William Parker Foulke, 18 March 1861 and 28 March 1861, William Parker Foulke Papers, APS. Also see "Dr. Hayes's Arctic Expedition," *Boston Journal,* 14 November 1860, 2; "The Hayes Arctic Expedition," *New-York Times,* 16 November 1860, 8.

15. Hayes, *Open Polar Sea,* 92, 108-109.

16. Ibid., 349, 378-379.

17. Ibid., 449-450; "The Arctic Expedition," *New-York Daily Times,* 11 October 1861, 4.

18. "An Arctic Boat Journey in the Autumn of 1854"; "Arrival of the Hayes Arctic Expedition," *Boston Daily Evening Transcript,* 23 October 1861, 2.

19. Grinnell to Foulke, 18 March 1861 and 28 March 1861, William Parker Foulke Papers, APS; "Arctic Explorations," *New-York Times,* 15 October 1861, 2; "Dr. Hayes' Exploration," 17-22. This account was originally published in the *North American* on 29 November 1861. Hayes lectured at the American Geographical Society, the American Philosophical Society, and the Academy of Natural Sciences of Philadelphia.

20. Caswell, *Arctic Frontiers,* 39; "The Arctic Expedition," *New-York Daily Times,* 11 October 1861, 4. The *Boston Daily Evening Transcript* gave Hayes higher marks for the expedition's scientific results: "Dr. Hayes's Arctic Expedition," 24 October 1862, 2; Isaac Hayes to Henry Grinnell, 23 October 1861, William Parker Foulke Papers, APS.

21. In the 1860s and 1870s, Kane's *Arctic Explorations* appeared frequently on lists of notable literature and recommended classroom literature. For a small sample of these, see *Free Public*

Libraries; Martin, *Choice Specimens,* 328-329; Northend, *Teacher's Assistant,* 345. School and public library catalogues from the same period show that *Arctic Explorations* had become widely available.

22. Hayes, *Open Polar Sea,* 226.

23. "Open Polar Sea," *Littell's Living Age;* "Open Polar Sea," *Atlantic Monthly;* Hayes, *Open Polar Sea,* 125.

24. With the help of Coast Survey geodesist Charles Schott, Hayes salvaged Sonntag's data and published them in Joseph Henry's scholarly series Smithsonian Contributions to Knowledge: Hayes, "Physical Observations in the Arctic Seas." "Scientific Matters," *Boston Daily Evening Transcript,* 13 July 1860, 1; W. P. Wheeler to Charles Hall, box 4, folder 57, Charles F. Hall Papers, Archive Center, National Museum of American History, Smithsonian Institution, Washington, DC (hereafter Hall Papers).

25. Not all of Hayes's interactions with the scientific community were so smooth. He quarreled with Henry about the publication of the data from the voyage. See letters from Joseph Henry to Isaac Hayes, 1861, Smithsonian Institutional Archives, Hayes collection. Hayes also alludes to these problems in the introduction of *Open Polar Sea.*

26. Hayes, "Transactions of the Society for 1868"; Hayes, "Transactions of the Society for 1870," lxxvi (quotation).

27. Loomis, *Weird and Tragic Shores,* 40-46; Hall, undated lecture, Hall Papers.

28. Hall, undated lecture, Hall Papers. Hall probably reached the number of 105 missing men by subtracting the men confirmed dead by Rae and McClintock from the original Franklin crew of 129.

29. Loomis, *Weird and Tragic Shores,* 60.

30. Sundquist, "Literature of Expansion and Race"; Rogin, *Fathers and Children;* Huhndorf, "Going Native," 3-7; Berkhofer, *White Man's Indian,* 86-104; Longfellow, "Song of Hiawatha," 114; Berkhofer, "White Conceptions of Indians."

31. Derounian-Stodola, *Indian Captivity Narrative,* 170.

32. Loomis, *Weird and Tragic Shores,* 47.

33. Ibid., 59.

34. "A New Arctic Expedition," *New-York Times,* 1 June 1860, 8.

35. On the search for the Northwest Passage, see Ross, "Nineteenth-Century Exploration of the Arctic," 244-331; Neatby, *In Quest of the Northwest Passage;* Fleming, *Barrow's Boys.*

36. Hall, *Arctic Researches,* iii.

37. Loomis, *Weird and Tragic Shores,* 86; Hall, *Arctic Researches.*

38. Untitled article, *New York Herald,* 6 November 1862; Loomis, *Weird and Tragic Shores,* 149-152.

39. C. L. Daboll to Charles Hall, 18 December 1862, Hall Papers; broadsheet for "Barnum's Museum," ibid.; Loomis, *Weird and Tragic Shores,* 153. Hall eventually grew concerned for the health of the Eskimo family and refused to extend the Barnum contract.

40. Loomis, *Weird and Tragic Shores,* 169-177.

41. Ibid., 177-229.

42. Joseph Henry to the National Academy of Science, reprinted in Davis, *North Polar Expedition,* 637-639.

43. Nourse, *Narrative,* 25; Alexander Dallas Bache to Charles Hall, 22 May 1863, Hall Papers; Nourse, *Narrative,* 37; Walker, "Captain Hall's Arctic Expedition," 203.

44. Walker, "Captain Hall's Arctic Expedition," 202; "Arctic Researches and Life Among the Esquimaux," 125. Similar criticism came from T. B. Maury, "Captain Hall's Arctic Expedition," 515.

45. T. Kane, *Alaska and the Polar Regions.*

46. Hall is quoted in a letter to A. B. Johnson, reprinted in C. Davis, *Narrative of the North Polar Expedition,* 19 (the loyalist, J. C. Brevoort, is quoted at 41). Selections from the *New York World* article were reprinted in "Arctic Researches." On Hall's testimony, see Loomis, *Weird and Tragic Shores.*

47. C. Davis, *Narrative of the North Polar Expedition*, 19; Loomis, *Weird and Tragic Shores*, 255. Hall lectured in many cities including New York City, Brooklyn, NY, Pittsburgh, PA, Cincinnati, OH, Indianapolis, IN, and Washington, DC. See C. Davis, *Narrative of the North Polar Expedition*, 21.

48. Nourse, *Narrative*, 33; Robeson, *Instructions for the Expedition*, 7; Walker, "Captain Hall's Arctic Expedition," 208.

49. Hall may have correctly diagnosed the cause of his illness. Loomis traveled to the Arctic and brought back a sample of Hall's remains. Chemical analysis reveals Hall's exposure to high levels of arsenic shortly before he died. Eugenia Vale Blake offers the most detailed account of the stranded party's incredible passage on the pack ice in *Arctic Experiences*. Hall's crewman is quoted by Loomis, *Weird and Tragic Shores*, 309.

50. Herbert, *Noose of Laurels*, 58. Herbert's critique of Hayes applies even more closely to Kane. In 1851, when Kane started campaigning for his own Arctic command, he had less Arctic experience than did Hayes.

CHAPTER FOUR

1. Two of the twenty-five members of the expedition, Jens Edward and Thorlip Frederick Christiansen, were native Greenlanders. The quotation is from Schley, *Rescue of Greely*, 222.

2. "The Rescue of Greely," *New York Herald*, 26 March 1885; Ford, "Heroes of the Icy North"; Brainard, *Outpost of the Lost*, 312; Brainard, *Six Came Back*, 301. "If we were Englishmen" comes from Charles Harlow, "Greely at Cape Sabine," *Century Magazine* (undated), in Adolphus Washington Greely Papers, Manuscript Division, Library of Congress, Washington, DC (hereafter Greely Papers). "If we've got to starve" is from Copley, "Will to Live," 502-503. "'Give us something to eat!'" comes from a newspaper clipping titled "Sergeant Connell" (source unknown), 31 May 1885, box 74, Greely Papers.

3. Dupree, *Science in the Federal Government*, 193; Barr, *Expeditions of the First International Polar Year*; Todd, *Abandoned*, 304.

4. Riffenburgh, *Myth of the Explorer*.

5. Henry Stanley, untitled article, *New York Herald*, 10 August 1872; Riffenburgh, *Myth of the Explorer*, 74-75.

6. Riffenburgh, *Myth of the Explorer*, 53, 72; Mott, *American Journalism*, 416; Guttridge, *Icebound*, 285.

7. On Bennett's congressional campaign, see Guttridge, *Icebound*, 32.

8. Ibid., 81-82, 100, 154-155.

9. Ibid., 176-179.

10. Editorial, *New York Times*, 20 January 1882, 4; Editorial, "De Long and His Men," *New York Times*, 22 March 1882, 4; Guttridge, *Icebound*, 275, 284-285; Editorial, "The Search for the Pole," *New York Times*, 24 December 1881, 4 (for the quotation); Dall, "Recent and Future Arctic Exploration," 378-379. See also Editorial, *Chicago Tribune*, 6 May 1882.

11. Editorial, "Siberian Perils-Melville's Report and De Long's Record," *New York Herald*, 21 March 1882, 8.

12. Guttridge, *Ghosts of Cape Sabine*, 111; Caswell, *Arctic Frontiers*, 41. The quotation is from Editorial, *New York Times*, 1 February 1881, 4.

13. George F. Edmunds is quoted in U.S. Congress, *Congressional Record*, 47th Cong., 1st sess. (1881), 1203; Dall, "Voyage of the Vega," 254-255.

14. Dall, "Our Lost Explorers (Review)," 402.

15. Barr, *First International Polar Year*, 3.

16. "Recent Polar Explorations," 320-332.

17. Hawes, "Signal Corps," 69.

18. Goetzmann, *Exploration and Empire*, 356. For background information, see Fleming, *Meteorology in America*.

19. Maury published many of his views of polar geography in *Putnam's Monthly Magazine*. For examples see "Gateways to the Pole"; "Dumb Guides to the Pole"; "Eastern Portal to the Pole." For his writings about Charles Hall, see "Captain Hall's Arctic Expedition."

20. House Committee on Naval Affairs, *Expedition to the Arctic Seas*, 46th Cong., 2nd sess., 1880, H. Rep. 453, 3. For similar testimonials see House Committee on Naval Affairs, *Expedition to the Arctic Seas*, 45th Cong., 2nd sess., 1878, H. Rep. 96, 10–11; Daly, "Annual Address of Chief Justice Daly" (1880); DeCosta, "Arctic Exploration."

21. Rebecca Robbins Raines, *Getting the Message Through*, 61.

22. Guttridge, *Ghosts of Cape Sabine*, 14–17; untitled article, *Frank Leslie's Illustrated Newspaper*, 7 May 1881, 1.

23. Guttridge, *Icebound*, 62.

24. Guttridge, *Ghosts of Cape Sabine*, 84, 89.

25. Ibid., 151–178, 269–277.

26. "Greely's Retreat," *New York Times*, 19 July 1884, 4; Editorial, *New York Times*, 20 July 1884, 6; Barr, *First International Polar Year*, introduction.

27. "Lieut. Greely Rescued," *New York Times*, 18 July 1884, 1; "Cheering Arctic Heroes," *New York Times*, 5 August 1884, 1–2; "Reception of the Greely Expedition Survivors at Portsmouth," *Frank Leslie's Illustrated Newspaper*, 16 August 1884, 401, 529. Henrietta later told her father that these accounts "were absurd." Henrietta Greely to Adolphus Greely, 3 August 1884, box 14, Greely Papers.

28. "Home from the Frozen Seas," *New York Times*, 2 August 1884, 1; "Reception of the Greely Expedition Survivors at Portsmouth," *Frank Leslie's Illustrated Newspaper*, 16 August 1884, 401; James Gordon Bennett to Adolphus Greely, 11 August 1884, box 14, Greely Papers; W. A. McGinley to Adolphus Greely, 7 August 1884, box 14, Greely Papers.

29. Eugene Hale is quoted in McGinley, *Reception of Lieut. A. W. Greely*, 35.

30. "Back from the Frozen Seas," 529.

31. "The Last of the Arctic Dead," *New York Times*, 10 August 1884, 3; W. B. Hazen to Adolphus Greely, 12 August 1884, box 14, Greely Papers.

32. "Horrors of Cape Sabine," *New York Times*, 12 August 1884, 1; "A Horrible Discovery," *New York Times*, 12 August 1884, 4.

33. Todd, *Abandoned*, 289–291; Guttridge, *Ghosts of Cape Sabine*, 298–299.

34. Greely collected these reports. See "A Sad Arctic Record" (source unknown), 7 February 1885, "Greely's Arctic Camp," *New York Post*, 7 February 1885, n.p., and "More Arctic Horrors" (source and date unknown), all in box 74, Greely Papers.

35. "Tale of the Diaries" (source unknown), box 74, Greely Papers.

36. W. B. Hazen to Adolphus Greely, 25 September 1884, box 14, Greely Papers.

37. Dwight, "Arctic Meeting," 338; Greely, "When I Stood Face to Face with Death," 2.

38. Untitled article, box 74, Greely Papers; untitled article, *Sun* (Pittsfield, MA), 223 July 1885; Greely, "The White North."

39. Blanchard, *Oscar Wilde's America;* "Explorer Greely as a Lady's Man," *Times* (Philadelphia), 25 January 1885; Catherine Cole, "Catherine Cole at the Capital," *Daily Picayune* (New Orleans), 14 March 1886, 12.

40. "Lieut. Greely Aping Oscar Wilde," *Buffalo Express*, n.d., box 74, Greely Papers. Blanchard considers Wilde's portrayal by the *Washington Post* (22 January 1882) and *Harper's Weekly* (28 January 1882) in *Oscar Wilde's America*, 32–33.

41. Savage, *Our Heroic Age*, 6:14; Melville, *In the Lena Delta*, 458.

42. Dwight, "Arctic Meeting," 341.

43. Greely, *Three Years of Arctic Service*, 2:337.

44. M. P., "The Capriciousness of Public Notice," *Earnest Worker* (March 1887), box 74, Greely Papers.

45. Roswell D. Hitchcock is quoted in Greely, "Arctic Meeting," 334.

46. Greely, "White North," 2; Raines, *Getting the Message Through*, 61.

47. "Schwatka's Search," *New York Times*, 11 December 1881, 6.

48. Greely, "Arctic Meeting," 331.

49. Ibid., 331-332.

50. Herzig, "In the Name of Truth," 76-113; "The Magnetic and Tidal Work"; Editorial, *Science* 4 (August 1884), 94; Gilman, "Reception of the Greely Arctic Explorer," 54.

51. "Smith Sound and Its Exploration," 623 (for Bessels quotation); H. Rink, "Polar Question," 691. Holt's letter of inquiry and scientists' responses to it are stored in the correspondence files of the American Geographical Society in New York. The quotations from the correspondence are taken from William Davis to Holt, 6 February 1894; Simon Newcomb to Holt, 9 February 1894; W. T. Sedgewick to Holt, undated (emphasis in original).

52. Untitled newspaper article, undated, Folder "LFB A," box 74, Greely Papers; Mill, "Race to the North Pole," 146; C. W. Darling to Adolphus Greely, box 14, Greely Papers; Editorial, *New York Daily Graphic*, 26 July 1884.

53. Schwatka, "Coming Polar Expeditions," 154.

54. De Long, *Voyage of the* Jeannette, 1.

55. M. P., "Public Notice," n.p. Hazen expressed similar sentiments in his correspondence with Greely. See W. B. Hazen to Adolphus Greely, 21 August 1884, box 14, Greely Papers.

CHAPTER FIVE

1. Wellman, *Aerial Age*, 24, 35. On the 1894 expedition, see Capelotti, *By Airship to the North Pole*, 60.

2. Wellman, *Aerial Age*, 16.

3. Peary, *North Pole*, 5.

4. Other histories of Wellman and Peary also consider their use of mechanical devices, but they focus on the function of these devices in the field. Edward Mabley's 1969 account *The Motor Balloon "America"* chronicles Wellman's Arctic voyages as preludes to his failed bid to cross the Atlantic Ocean in 1910. *By Airship to the North Pole* by P. J. Capelotti considers artifacts from *America*'s Arctic base camp with an eye toward resolving questions about Wellman's polar campaign. Mabley and Capelotti both acknowledge the role of Wellman's airships in generating publicity, using this information to shed light on Wellman's personal motives rather than on broader cultural currents. In a similar manner, histories of Peary such as Robert M. Bryce's comprehensive *Cook and Peary* generally consider a modern innovation, such as Peary's specially designed ship *Roosevelt*, for its performance in the Arctic rather than for its symbolic value back home. Also see William Herbert Hobbs's partisan biography *Peary* for useful discussion of Peary's assimilation of Eskimo equipment and travel techniques. Although scholars have paid little attention to the role of exploration equipment in popular campaigns, they have amply demonstrated the ways explorers' attitudes about mechanization influenced their expeditions. In particular, they have shown how explorers often adopted and discarded polar equipment on the basis of their cultural prejudices. See Roland Huntford's *The Last Place on Earth* and Barry Pegg's "Nature and Nation."

5. In looking at cultural debates through the lens of machines used in the Arctic, I have been guided by an extensive body of scholarship on technology and gender symbolism. Joan Scott argues in "Gender" that gender impacts the story of machines at many different levels. Mechanical devices reflect the cultural attitudes of not only the engineers who design them but also the consumers who use them. Standing between these groups are advertisers who promote these devices and advise consumers about their proper function. I have chosen to focus on the promotion of machines used in the Arctic because it is at this level that ideals of gender become most visible. Wellman and Peary did not publicize their expeditionary machines in hopes that Americans would buy them. Rather,

they manipulated the symbolism of such machines as a means of self-promotion, using it to shape their own public images as explorers. In effect, these machines offered them a new way to market themselves, a strategy already used to good effect by inventors such as Robert Fulton, Thomas Edison, and Alexander Graham Bell, who drew on the symbolism of machines to elevate their own manly reputations. For more on the manly image of engineers, see Ruth Schwartz Cowan's *Social History of American Technology*, 208–212.

6. For more on the gender symbolism and technology, see Oldenziel, *Making Technology Masculine*, 19, 31; Taylor, "Register of the Repressed"; Caputi, "Metaphors of Radiation," 423–42; Cockburn, *Brothers*.

7. Seven men had died on previous American expeditions to the high Arctic: three on the second Grinnell expedition, undertaken by Kane, two on the Hayes expedition of 1860–1861, one on the Hall expedition of 1864–1868, and one on the Hall expedition to the North Pole of 1870–1871.

8. Markham, "North Polar Problem"; Mayo, "Ice-Breaker 'Ermack'"; "Another Polar Failure," 376; "To the North Pole by Automobile," 805; "Electric Light for Polar Explorers"; "Tires for Arctic Use"; Greely, "Nansen's Polar Expedition."

9. The plan for an Arctic railroad comes from an anonymous letter to the American Geographical Society, General Correspondence, AGS, New York. Peary, *North Pole*, 17–18. Peary was not exaggerating about the incredible number of schemes. Every year he received dozens of letters about inventions. The numbers increased during years when he was planning a new expedition. For examples, see Hiram Maine to Peary, 11 April 1907; D. F. Akin to Peary, 10 April 1907; Nathaniel Ladd to Peary, 15 February 1907; William Lees to Peary, 15 April 1907; Gavin Learmonth to Peary, 4 March 1907; Charles Fiesse to Peary, 19 March, 1907; J. H. Emslie to Peary, 15 April 1907. All are in the Robert E. Peary Papers, National Archives, College Park, MD (hereafter Peary Papers). Other explorers received similar letters; see 1909 letters to Frederick Cook, Frederick A. Cook Papers, Manuscript Division, Library of Congress, Washington, DC. David E. Nye examines the rising power of technologies as cultural symbols in the nineteenth century in *American Technological Sublime*, 24–25.

10. On Wellman's early years, see "Walter Wellman, Explorer, Is Dead," *Newark News*, 1 February 1934, in Wellman's biographical file at the National Museum of Air and Space, Smithsonian Institution, Washington, DC; Mott, *American Journalism*, 460; Mabley, *Motor Ballooon "America,"* 16–17. On the rise of globe-trotting reporters such as Stanley, see Riffenburgh, *Myth of the Explorer*; Mott, *American Journalism*, 415–421. On the involvement of Lawson and the *Chicago Record-Herald* in expeditions, see Wellman, *Aerial Age*, 128.

11. Articles by Wellman include "On the Way to the Pole" (quotation is at 535); "Race for the North Pole"; "Sledging Toward the Pole"; "Arctic Day and Night"; "Long-Distance Balloon Racing"; "Polar Airship"; "By Airship to the North Pole."

12. Hayes, *Open Polar Sea*, 216–217.

13. Wellman, "Arctic Day and Night," 562. He reprinted this passage later in *Aerial Age*, 63.

14. Mabley, *Motor Balloon "America,"* 20.

15. La Vaulx is quoted by Wellman in "Long-Distance Balloon Racing," 212.

16. On early interest in balloon flight, see Tucker, "Voyages of Discovery"; Crouch, *Eagle*; DeVorkin, *Race to the Stratosphere*. The information about Kane and balloon flight is from Robert Mills to E. K. Kane, 27 December 1852; John Wise to Robert Mills, 13 January 1853; R. F. Lewis to E. K. Kane, 28 March 1853. All are in the Elisha Kent Kane Papers, American Philosophical Society Library, Philadelphia. On De Long's consideration of balloon flight, see De Long, *Voyage of the Jeannette*, 49–50. On Cheyne, see "The Search for the Pole," *New York Times*, 24 December 1881, 4; "A New Arctic Expedition," *New York Times*, 9 February 1880, 3. The quotation is from Prentiss, *Great Polar Current*, 28. In 1930, whalers found the skeletons of Andrée and his crew on White Island, northeast of Spitsbergen.

17. Mabley, *Motor Balloon "America,"* 21.

18. The *Providence Journal* is quoted in an untitled article in the *Chicago Record-Herald*, 11 January 1906, 5; the *Intelligencer* (Wheeling, WV) is quoted in "Government to Aid in North Pole Hunt," *Chicago Record-Herald*, 4 January 1906, 5.

19. "Wellman's Voyage," *Chicago Record-Herald*, 1 January 1906, 1-2.

20. The newspapers are quoted in "Praise for Pole Hunt," *Chicago Record-Herald*, 3 January 1906, 4; "Praise and 'Knocks' by Press," *Chicago Record-Herald*, 5 January 1906, 4. The Wellman quotation comes from *Aerial Age*, 157.

21. Wellman, "By Airship to the North Pole," 199-200.

22. Capelotti, *Airship*, 128.

23. The sealing ship *Farm* should not be confused with Fridtjof Nansen's expedition vessel *Fram*. On the failure of his expedition and the rescue by *Fram*, see Wellman, *Aerial Age*, 194.

24. Peary, *Nearest the Pole*, x; Hobbs, *Peary*, 43; Bryce, *Cook and Peary*, 19-20.

25. Peary, "Diary," in *Notebooks, Diaries, Journals, 1871-1886*, Peary Papers; Hobbs, *Peary*, 72.

26. Bryce, *Cook and Peary*, 65; Huntford, *Nansen*, 100-116. Peary introduced other mechanical innovations in his later attempts on the poles. He designed a new sledge, dubbed the Peary Sledge, that was longer and supposedly more sturdy than traditional Inuit sledges. He also designed an alcohol stove that could heat water more quickly than those using earlier designs, thereby saving time on his final polar dash. Peary, *North Pole*, 9.

27. Bryce, *Cook and Peary*, 44-45, 65-66. Peary was slow, however, to appreciate the benefits of igloos; see Hobbs, *Peary*, 139.

28. Peary, *Northward Over the "Great Ice,"* 1:lxii; Dyche, "Curious Race of Arctic Highlanders," 235; Hobbs, *Peary*, 124. The quotation is from Peary, *Secrets of Polar Travel*, 197, quoted in Pegg, "Nature and Nation," 221.

29. For more on Schwatka's Arctic career see Gilder, *Schwatka's Search;* "Lieut. Schwatka's Search," *New York Times*, 26 September 1880, 6. Also see Schwatka's "Letter of Lieut. Frederick Schwatka"; "Alaska Military Reconnaissance of 1883"; "Arctic Vessel and Her Equipment"; "Icebergs and Ice-floes"; "Wintering in the Arctic'"; "Exploration of the Yukon River." The quotation is from Daly, "Arctic Meeting at Chickering Hall," 260.

30. Loomis, *Weird and Tragic Shores*, 86-155.

31. See the section concerning Charles Hall in chapter 3.

32. U.S. Bureau of the Census, *Historical Statistics*, 1:8; Mills is quoted in Machor, *Pastoral Cities*, 121.

33. Rosenberg, *Cholera Years*. Chapin is quoted in Machor, *Pastoral Cities*, 149.

34. U.S. Bureau of the Census, *Historical Statistics*, 1:11; Sicherman, "Uses of a Diagnosis." On fears of immigration and race suicide, see Bederman, *Manliness and Civilization;* Nash, "American Cult of the Primitive"; Wells, *War of the Worlds;* Wells, *Time Machine*.

35. Amory Mayo is quoted in Machor, *Pastoral Cities*, 122.

36. On G. Stanley Hall, see Bederman, *Manliness and Civilization;* Huhndorf, "Going Native," 54-72.

37. Turner, *Significance of the Frontier*. See Fabian, "Ragged Edge of History." On Hall, see Bederman, *Manliness and Civilization*. Examples of "savage" novels include Edgar Rice Burroughs, *Tarzan of the Apes* (Chicago: A. C. McClurg, 1914); Jack London, "Call of the Wild," *Saturday Evening Post*, June 20-July 18, 1903; Joseph Conrad, *Lord Jim* (Garden City, NY: Doubleday, 1899); and Conrad, *The Heart of Darkness* (Edinburgh: W. Blackwood, 1899). The quotation is from F. Cooper, "Primordialism and Some Recent Books," 278.

38. Morris, *Rise of Theodore Roosevelt;* Bederman, *Manliness and Civilization*.

39. U.S. Bureau of the Census, *Historical Statistics*, 1:11, 22, 32; Bryce, *Cook and Peary*, 324. For information on Peary's extensive involvement in the club, see the following letters from club organizers in "Letters Received," Peary Papers: Forest Fish and Game Society of America to Peary,

14 December 1907, and New England Forest Fish and Game Association to Peary, 22 March 1907. In same collection, also see letters in "Social and Professional Organizations" such as American Alpine Club to Peary, undated.

40. Harry Radford to Robert Peary, 4 September 1907, Peary Papers.

41. Hellman, *Bankers, Bones, and Beetles*, 85; Hobbs, *Peary*, 177; Bryce, *Cook and Peary*, 132-133.

42. Bryce, *Cook and Peary*, 210-212.

43. Peary, "Gold Medal," 693-694.

44. The *Marine Review* in quoted in "Peary's New Arctic Ship 'Roosevelt,'" 780-781.

45. Peary, *Nearest the Pole*, 134; Bryce, *Cook and Peary*, 289. Peary's 1905 record for highest latitude is not without controversy. See Herbert, *Noose of Laurels*, 273; Bryce, *Cook and Peary*, 854.

46. Pauly, "The World and All That Is in It."

47. Huntford, *Nansen*, 150.

48. Gilbert Grosvenor to Robert Peary, 15 January 1907, Peary Papers; Bryce, *Cook and Peary*, 291. Theodore Roosevelt is quoted in Peary, *Nearest the Pole*, vii.

49. Peary, *Nearest the Pole*, 44-45.

50. Peary, *North Pole*, 35.

51. Bryce, *Cook and Peary*, 418.

52. Wellman, *Aerial Age*, 9.

53. Peary, *North Pole*, 121.

CHAPTER SIX

1. Bryce, *Cook and Peary*, 391. For these accounts of pro-Cook activities see Clarence Brigman to Frederick Cook, 22 September 1909, Harry Bush to Frederick Cook, 22 September 1909, and Anti-tuberculosis League to Frederick Cook, 9 September 1909, all in reel 2 of the Frederick Albert Cook Papers, Library of Congress, Manuscript Division, Washington, D.C. (hereafter Cook Papers).

2. White men traveled with Cook and Peary into the Arctic, but none accompanied the explorers on the final leg of their polar journeys.

3. There is a vast body of literature about the events of 1909. The most comprehensive account is Bryce's *Cook and Peary*. Lisa Bloom's *Gender on Ice* analyzes the symbols and rhetoric of Arctic narratives to see how they function as cultural texts. But Bloom looks at these texts in isolation and in so doing overestimates their persuasiveness with popular audiences and their representativeness of ideals in American culture. Focusing on a small number of texts, one can see why Bloom concludes that Peary emerged as "an epitome of manliness," ending his career as a national hero. But when we look at the broad range of texts about the controversy, from letters and editorials to cartoons, we can see that the explorers remained controversial figures, not only in their claims of discovery but in the way they portrayed themselves as men.

4. "Sir John Franklin Expedition," 890; Davis, Gallman, and Gleiter, *In Pursuit of Leviathan*, 39. For a good example of explorers using whaling arguments to gain support, see Charles Hall to Edward Everett, 16 April 1863, folder 46, Charles Francis Hall Collection, Smithsonian Institution, Washington, DC. Among the more imaginative commercial proposals put forward by supporters was a "house of entertainment" at the North Pole for wealthy tourists. See Sonntag, *Professor Sonntag's Thrilling Narrative*.

5. "Results of the Arctic Search"; House Committee on Naval Affairs, *Expedition to the Arctic Seas* (46th Cong.), testimony of Louis Agassiz; Davis, Gallman, and Gleiter, *In Pursuit of Leviathan*, 7. By the last decade of the nineteenth century, commercial gains from Arctic exploration appeared even less likely. The whaling industry had become a shadow of its former self, with annual receipts declining to 14 percent of those collected in the 1850s. The promise of a trade route across the polar sea had diminished as well. De Long's and Nansen's long periods of drifting in its pack ice effectively put to rest any hope of finding an open passage to Asia across the top of the world. Even

if the uncharted regions of the polar sea had revealed an open channel, planning was already under way for a transoceanic canal across the isthmus of Panama.

6. Rubin, *Making of Middlebrow Culture*, 2-10; Cochran, *200 Years of American Business*, 8. William Leach uses the phrase "pecuniary values" in *Land of Desire*, 7.

7. Greely, "Scope and Value of Arctic Explorations," 34.

8. Wellman, *Aerial Age*, 10.

9. Greely's statement and record of contracts come from the Adolphus Washington Greely Papers, Manuscript Division, Library of Congress, Washington, DC. The quotation from Adolphus Greely to the *New York Times*, 1 January 1924, box 73. On contracts, see Harper Brothers to Greely, 18 July 1884, *Century Magazine* to Greely, 18 September 1884, Thorndike Rice (*North American Review*) to Greely, 15 December 1884 and 22 December 1884, James Parton (*Youth Companion*) to Greely, 28 November 1884, Edgar Wakeman to Greely, 10 November 1884 and 6 December 1884, D. W. Leslie to Greely, 22 July 1884 and 24 July 1884, James B. Pond to Greely, 31 August 1884, J. M. Stoddart to Greely, 18 July 1884, and R. W. Smith (*Scribner's*) to Greely, 1 November 1884, all in box 14, Greely Papers. In 1888, Greely published *Three Years in Arctic Service*. For contracts involving his men, see, e.g., David Brainard to Greely, 26 August, 1884, 22 September 1884, and 30 September 1884, Greely Papers. Also see Guttridge, *Ghosts of Cape Sabine*, 298-300.

10. Peary, "Gold Medal of the Paris Geographical Society," 694.

11. Peary, *Nearest the Pole*, ix.

12. Ford, "Heroes of the Icy North," 288; Sadie E. Bigelow to Robert Peary, 2 May 1907, box 30, Robert E. Peary Papers, Rear Admiral Robert E. Peary Family Collection, National Archives, College Park, MD (hereafter Peary Papers); Peary, "Peary's Work in 1901-1902," 386.

13. Constance Goddard DuBois to Robert Peary, 17 April 1907, box 31, Peary Papers.

14. Hobbs, *Peary*, 123; Bryce, *Cook and Peary*, 1024; Peary, *Nearest the Pole*, 355. Eventually, Peary sold the meteorites to the AMNH for $40,000. On Peary's role in bringing the six Eskimos to New York, see Harper's *Give Me My Father's Body*.

15. For examples, see DuBois to Peary, 15 April 1907, and M. Jones to Peary, 6 April 1907, Peary Papers. One only has to look at a map of northern Greenland and Ellesmere Island to see the names of Peary's most generous supporters. Wally Herbert outlines Peary's ambitious plans for postvoyage self-promotion in *Noose of Laurels*.

16. Bryce, *Cook and Peary*, 3-6.

17. Cook, "The Most Northern Tribe on Earth," *New York Medical Examiner* 3 (1893): 23-24, quoted in Bryce, *Cook and Peary*, 94.

18. Ibid., 58-59, 105-106.

19. Ibid., 111; "Cook Gave a Show in a Dime Museum," *New York Times*, 10 September 1909, 4.

20. Cook's published writings about these expeditions include *Through the First Antarctic Night, 1898-1899: A Narrative of the Voyage of the "Belgica" Among Newly Discovered Lands and Over an Unknown Sea About the South Pole* (New York: Doubleday and McClure, 1900); "America's Unconquered Mountain," *Harper's Monthly Magazine*, January 1904, 230-239; *To the Top of the Continent: Discovery, Exploration, and Adventure in Sub-Arctic Alaska; The First Ascent of Mt. McKinley, 1903-1906* (New York: Doubleday, Page, 1908).

21. Bryce, *Cook and Peary*, 294-295, 349.

22. Cook, "Arctic Regions as a Summer Resort," 264.

23. Dunn, *The Shameless Diary of an Explorer* (New York: Outing Publishing Co., 1907), quoted in Bryce, *Cook and Peary*, xviii.

24. "Dr. Cook Wins Friendship of Great Audience That Assembles at Carnegie Hall to Hear His Story," *New York Herald*, 28 September 1909, 3.

25. Anthony Fiala, "Smooth Ice Helped, Thinks Mr. Fiala," *New York Herald*, 3 September 1909, 4; Greely, introduction to *Discovery of the North Pole*, n.p.

26. Frederick Cook, as recalled by Philip Gibbs in "Adventures of an International Reporter," *World's Work* 45 (March 1923): 481, quoted in Bryce, *Cook and Peary*, 366; untitled article, *New York Herald*, 8 September 1909.

27. Adams, "North Pole at Last," 422; Greely, "Discoverers of the North Pole."

28. For discussion of Eskimo witnesses at the North Pole see "The 'Big Nail' and Little Hammers," 513. On forms of proof, see Reid, "How Could an Explorer Find the Pole?"; Bryce, *Cook and Peary*, 1099-1100.

29. Walter Wellman is quoted in Bryce, *Cook and Peary*, 860-861.

30. Quoted in "Polar Perpetrations."

31. C. M. Coomer to Cook, 22 September, 1909, reel 2; Ira Kennion Wood to Cook, [n.d.], reel 3; I. Dockery to Cook, 17 October 1909, reel 2; Paul Nesbit to Cook, 22 September 1909, reel 3; H. Fischer to Cook, 14 September 1909, reel 2, all in Cook Papers. The New York City ceremony for Cook, reported in the *New York Herald*, 16 October 1909, is quoted in Bryce, *Cook and Peary*, 434.

32. "Scientific Proofs" is quoted in "Polar Perpetrations," 605.

33. The cartoon is from Miller, *Discovery*, preface; Cook is quoted in "The 'Big Nail' and Little Hammers," 512. Greely, "Discoverers of the North Pole," 290.

34. E. Wyckoff to Cook, Cook Papers. The Cook Papers are full of correspondence from publishers eager for the rights to Cook's North Pole narrative. See reels 2 and 3 in particular. For information on Cook's lecture tours, see Bryce, *Cook and Peary*, 352, 374, 420-423, 437.

35. *New York Times* and *New York Post* articles are quoted in "Peary's 'Proofs,'" 661. The cartoons, "The Arduous Journey" (from *Literary Digest* [September 1909]: 466; "From Greenland's Icy Mountains," *Literary Digest* [October 1909]: 711; Minor, "A Profitable Mill," *St. Louis Post-Dispatch*, n.d.) are reprinted in "Peary's 'Proofs,'" 661. Peary is quoted in Bryce, *Cook and Peary*, 440.

36. Bryce, *Cook and Peary*, 428-30; "Peary's 'Proofs,'" 659.

37. "Dr. Cook Lived Like a Savage in Order to Gain the Pole," *New York Herald*, 18 September 1909, 3-4.

38. Stead is quoted in Miller, *Discovery*, 64; Gibbs is quoted in Bryce, *Cook and Peary*, 357; Steensby is quoted in "A Question of the Poles," 736-737; "Press Opinions of Dr. Cook and the Herald," *New York Herald*, 4 September 1909, 6; Mary Brennan, one of the dozens of admirers who sent Cook poems praising his accomplishment, wrote, "He sat not where there was comfort / Nor in comfort did recline / But, this man who loved to travel went / Exploring, and thought it was fine." Brennan to Cook, 22 September 1909, reel 2, Cook Papers.

39. "To the North Pole Along the Milky Way," *New York Herald*, 4 September 1909, 2.

40. Gilbert Grosvenor to Robert Peary, 18 January 1907, Peary Papers; "bearish" is from the *Camden Post-Telegram*, quoted in "Editors Condemn the Attitude of Mr. Peary in the Polar Battle," *New York Herald*, 1 October 1909, 4; "savage dominance" is from the *New York World*, 21 September 1909, quoted in Bryce, *Cook and Peary*, 394; "snarl" is from the *New York World*, 22 September 1909, quoted in ibid.; "ungracious" is from the *Savannah Mercury News*, quoted in "Editors Condemn the Attitude of Mr. Peary."

41. "'Dr. Cook a Man, Mr. Peary a Child,'" *New York Herald*, 10 September 1909, 7 (for Sverdrup); "has been childish" quoted in the *New York Herald*, 2 October 1909, 4; "American in London Likens Peary to a Baby That Has Lost Its Candy," *New York Herald*, 11 September 1909, 1; "baby act" is quoted in the *New York Herald*, 2 October 1909, 4; E. H. Ross, "Cook vs. Peary," *Pittsburgh Independent*, 25 September 1909.

42. Harper, "As the Matter Stands," first appears in the *New York American* and is reprinted in *Literary Digest* 39 (September 1909): 466; "The Foundling" is reprinted in *American Review of Reviews* 40 (October 1909): 416; "More Annexation Troubles" is reprinted in "The 'Big Nail' and Little Hammers."

43. Galton, "Heredity Talent and Character."

44. John Johnson Jr., "The Savagery of Boyhood," 799; Pearson, foreword to *Life, Letters, and Labors of Francis Galton;* Roosevelt, "American Boy," 2. Also see Lears, *No Place of Grace,* 144-147.

45. The Report of the University of Copenhagen Commission is quoted in Bryce, *Cook and Peary,* 470.

46. Robert Peary to R. A. Bartlett, 6 December 1910, Peary Papers.

47. Brennan to Cook, 22 September 1909, Cook Papers; Peck, "Disputes of Great Discoverers," 496; Balch, *North Pole and Bradley Land,* 15.

48. Depew is quoted in Bryce, *Cook and Peary,* 473, 1046.

CONCLUSION

1. Frederick A. Cook, "The Vibrating Aboma of Hate about the Pole," reel 8:83-91, n.d., Frederick Albert Cook Papers, Manuscript Division, Library of Congress, Washington, DC (hereafter Cook Papers).

2. *Nature* is quoted in "Science and Polar 'Dashes'"; Franz Boas, draft of an article later published in the *Independent* titled "Polar Exploration, Peary and Cook," 2-3, Franz Boas Papers, American Philosophical Society Library, Philadelphia; Louis Agassiz to Mr. LeConte, 6 December 1860, LeConte Papers, American Philosophical Society, Philadelphia.

3. On the evolution of U.S. geographical societies, see Wright, *Geography in the Making;* James and Martin, *Association of American Geographers;* Pauly, "The World and All That Is in It"; Schulten, *Geographical Imagination,* 49-50. Although many geographers rejected the work of amateur explorers, they nonetheless recognized the value of popular geography to the nation. Ibid., 52-53. The quotation is from Davis, "Need of Geography," 157.

4. Hevly, "Heroic Science."

5. Rotundo, *American Manhood,* 225; Peary, *North Pole,* 141.

6. Peary, *Northward Over the "Great Ice,"* 1:38; De Long, *Voyage of the* Jeannette, 21; Cook, *Frederick Albert Cook,* Cook Papers; Mill, "Race to the North Pole," 146.

7. That Byrd reached the North Pole remains a matter of dispute.

8. Ross, "Nineteenth-Century Exploration of the Arctic," 298.

9. This view is less dominant within exploration narratives of the Far West, perhaps because these expeditions set out with an abundance of practical objectives.

10. Anderson, *Ulysses Factor,* 20.

11. George W. Bush, quoted in National Aeronautics and Space Administration, "Vision for Space Exploration." Proponents of human space flight often claim that humans are more versatile than robots and better able to improvise. This argument carries less weight now that robots can be reprogrammed remotely to deal with unexpected problems and opportunities.

Bibliography

UNPUBLISHED PRIMARY SOURCES

American Geographical Society (New York, NY)
General Correspondence

American Philosophical Society (Philadelphia, PA)
Franz Boas Papers
William Parker Foulke Papers
Elisha Kent Kane Papers
LeConte Family Papers

Dartmouth College (Hanover, NH)
Vilhjalmur Stefansson Collection (Rauner Special Collections Library)

Library of Congress (Washington, DC)
Papers of Frederick Cook
Papers of Adolphus Greely

National Archives (College Park, MD, and Washington, DC)
Journal of Robert Carter (Record Group 45)
Papers of Emma De Long (Record Group 401)
Papers of George Washington De Long (Record Group 401)
Rear Admiral Robert Peary Family Collection (Record Group 401)

Smithsonian Institution (Washington, DC)
Spencer F. Baird Papers (Smithsonian Institutional Archives)
William Bradford Papers (Archives of American Art)
Sam DeVincent Collection of Illustrated Sheet Music (National Museum of American History)
Charles Hall Papers (National Museum of American History)
James Hamilton Scrapfile (National Museum of American Art)
Joseph Henry Papers (Smithsonian Institutional Archives)
Walter Wellman Scrapfile (National Museum of Air and Space)

United States Naval Academy (Annapolis, MD)
Journal of Lt. John C. Colwell (Special Collections and Archives Division, Nimitz Library)

University of New England (Portland, ME)
Josephine Peary Papers (Maine Women Writers Collection)

PUBLISHED PRIMARY SOURCES

Adams, Cyrus. "The Farthest North: An Account of Dr. Nansen's Adventures and Achievements." *McClure's Magazine*, December 1896, 99–109.
———. "The North Pole at Last." *American Review of Reviews*, October 1909, 422.
Alger, William R. *A Brief Tribute to the Life and Character of Dr. Kane.* Boston: A. Williams, 1857.
Allen, Joseph Henry. "Elisha Kent Kane: A Discourse Delivered at the Union Street Church, Bangor, Sunday, March 1, 1857." N.p.: 1857.
"The American Arctic Expedition." *Harper's New Monthly Magazine*, December 1851, 11–22.
Andrews, E. W. *Memoir and Eulogy of Dr. Elisha Kent Kane.* Boston: Dexter and Brother, 1857.
"Another Polar Failure." *Literary Digest*, 28 September 1901, 376.
"An Arctic Boat Journey." *Littell's Living Age*, July 1860, 150–151, 190–192.
"An Arctic Boat Journey in the Autumn of 1854." *Littell's Living Age*, June 1860, 792–794.
"Arctic Contributions to Science." *Littell's Living Age*, January 1853, 231–234.
"The Arctic Expedition." *Littell's Living Age*, March 1850, 453–456.
"The Arctic Expedition." *Catholic World*, November 1865, 279–280.
"Arctic Researches and Life among the Esquimaux." *Atlantic Monthly*, July 1865, 125.
Arctic Travels; or, An Account of the Several Land Expeditions to Determine the Geography of the Northern Part of the American Continent. New York: Carlton and Lanahan, n.d.
Astrup, Eivind. "In the Land of the Northernmost Eskimo." *Littell's Living Age*, April 1896, 102–114.
"The Automobile in Polar Exploration." *Literary Digest*, 8 October 1904, 454–455.
B. "A Gleam of the Northern Light." *Littell's Living Age*, November 1854, 193.
Bache, Alexander Dallas. "Abstract of the Principal Results of the Magnetic Observations of the Second Grinnell Expedition, in 1853–5, at Rensselaer Harbor and Other Points on the West Coast of Greenland, by Elisha K. Kane." *Proceedings of the American Association for the Advancement of Science* 12 (1858): 120–121.
Back, George. *Narrative of the Arctic Land Expedition to the Mouth of the Great Fish River and along the Shores of the Arctic Ocean in the Years 1833, 1834, and 1835.* 2d ed. Philadelphia: E. L. Carey and A. Hart, 1837.
"Back from the Frozen Seas." *Harper's Weekly*, 16 August 1884.
Baker, Edward. "The Franklin Expedition." *Congressional Globe* 19 (April 1850): 834.
Balch, Edwin Swift. "The Highest Mountain Ascent." *Journal of the American Geographical Society of New York* 36 (1904): 107–109.
———. *The North Pole and Bradley Land.* Philadelphia: Campion, 1913.
Barrington, Daines. *The Possibility of Approaching the North Pole.* New York: James Eastburn, 1818.
Bayly, Thomas. "Franklin Expedition." *Congressional Globe* 19 (April 1850): 830–835.
Bent, Silas. *An Address Delivered before the St. Louis Mercantile Library Association, January 6th, 1872, upon the Thermal Paths to the Pole, the Currents of the Ocean, and the Influence of the Latter upon the Climates of the World.* St. Louis: R. P. Studley, 1872.
"The 'Big Nail' and Little Hammers." *Literary Digest*, 2 October 1909, 512–514.
Bird, Robert Montgomery. "This Ice-Island." *Philadelphia Monthly Magazine*, December 1827, 113.

Blake, Eugenia Vale. *Arctic Experiences: Containing Capt. George E. Tyson's Wonderful Drift on the Ice-floe, a History of the Polaris Expedition, the Cruise of the Tigress, and the Rescue of the Polaris Survivors; To Which Is Added a General Arctic Chronology.* New York: Harper and Brothers, 1874.

Boas, Franz. "The History of Anthropology." In *The Shaping of American Anthropology, 1883–1911: A Franz Boas Reader,* edited by George W. Stocking, 23–36. New York: Basic Books, 1974.

———. "A Journey in Cumberland Sound and on the West Shore of Davis Strait in 1883 and 1884." *Journal of the American Geographical Society of New York* 16 (1884): 242–272.

———. "Polar Exploration, Peary and Cook." *Independent,* [November?] 1909, Franz Boas Papers, American Philosophical Society, Philadelphia.

Bogart, Fred R. "Those North Polers." *Bulletin,* September 1909, Frederick A. Cook Papers, Library of Congress, Washington, DC.

Bragg, Benjamin. *A Voyage to the North Pole.* London: G. Walker, 1817.

Brainard, David L. *The Outpost of the Lost: An Arctic Adventure.* Indianapolis: Bobbs-Merrill, 1929.

———. *Six Came Back: The Arctic Adventures of David L. Brainard.* Indianapolis: Bobbs-Merrill, 1940.

Breed, W. P. *Anthropos.* Philadelphia: Presbyterian Board of Publication, 1865.

Brinkerhoff. "From Greenland's Icy Mountains." *Literary Digest,* 30 October 1909, 711.

Brinton, Daniel Garrison. *The Myths of the New World: A Treatise on the Symbolism and Mythology of the Red Race of America.* New York: H. Holt, 1876.

Bryant, William C. "The Arctic Lover to His Mistress." *Lady's Book,* September 1833, 160.

Butler, Arthur P. "The Sir John Franklin Expedition." *Congressional Globe* 19 (May 1850): 891.

A Centennial Edition of the History of the United States: From the Discovery of America, to the End of the First One Hundred Years of American Independence. Washington, DC: Blair and Rives, 1875.

Chamberlain, Nathan H. *Dr. Kane and Christian Heroism as Seen in Arctic Voyaging.* Boston: Crosby, Nichols, 1857.

Champ, W. S. "The Ziegler Polar Expedition." *National Geographic,* October 1905, 427–428.

Clark, P. "The Symmes Theory of the Earth." *Atlantic Monthly,* April 1873, 471–480.

Clay, Henry. "Search for Sir John Franklin." *Congressional Globe* 19 (April 1850): 644.

Cole, Thomas. "Essay on American Scenery." *American Monthly Magazine,* January 1836, 1–12.

Cook, Frederick. "The Arctic Regions as a Summer Resort." *Home and Country,* October 1894, 257–264.

———. "The Conquest of Mount McKinley." *Harper's Monthly Magazine,* May 1907, 821–837.

———. "Factors in the Destruction of Primitive Man." *Brooklyn Medical Journal* 18 (September 1904): 333–335.

———. "The Greenlanders." In *The Last Cruise of the* Miranda, edited by Henry Collins Walsh. New York: Transatlantic, 1895, 172–177.

Cooper, Frederic Taber. "Primordialism and Some Recent Books." *Bookman* 30 (November 1909): 278–282.

Cooper, James Fenimore. *The Pioneers, or the Sources of the Susquehanna.* London: H. Colburn and R. Bentley, 1832.

———. *The Sea Lions.* New York: Stringer and Townsend, 1849.

Copley, Frank B. "The Will to Live." *American Magazine,* February 1911, 494–503.

Dall, William Healey. "The Dutch in the Arctic Seas." *Nation,* 28 March 1878, 216–217.

———. "Ice-Pack and Tundra," *Nation,* 15 March 1883, 240–241.

———. "Nils Adolf Erik Nordenskiold." *Nation,* 25 December 1879, 441–442.

———. "Our Lost Explorers (Review)." *Nation,* 9 November 1882, 402.

———. "Recent and Future Arctic Exploration." *Nation,* 1 November 1883, 378–379.

———. "Recent Arctic Literature." *Nation,* 30 October 1879, 296–297.

———. "Report of the Cruise of the United States Revenue Steamer 'Corwin' in the Arctic Ocean." *Nation*, 28 April 1881, 304–305.

———. "Schwatka's Search." *Nation*, 24 November 1881, 420.

———. "Sir John Franklin." *Nation*, 13 October 1881, 300.

———. "The So-Called ' *Jeannette* Relics.'" *National Geographic*, March 1896, 93–98.

———. "The Voyage of the Vega." *Nation*, 23 March 1882, 254–255.

Daly, Charles P. "Annual Address." *Journal of the American Geographical (and Statistical) Society* 2 (1870): lxxxiii–cxxvi.

———. "Annual Address of Chief Justice Daly." *Journal of the American Geographical Society of New York* 10 (1878): 1–76.

———. "Annual Address of Chief Justice Daly." *Journal of the American Geographical Society of New York* 12 (1880): 1–103.

———. "Arctic Meeting at Chickering Hall, October 28th, 1880." *Journal of the American Geographical Society of New York* 12 (1880): 237–296.

Dana, James D. "Arctic Explorations, by Elisha Kent Kane, U.S.N." *American Journal of Arts and Sciences* 24 (1857): 235–251.

Danenhower, John. "The Polar Question." *Proceedings of the United States Naval Institute* 11 (1885): 633–699.

Davis, C. H. *Narrative of the North Polar Expedition.* Washington, DC: Government Printing Office, 1876.

Davis, William Morris. "The Need of Geography in the University." *Educational Review* 10 (1895): 22–41.

DeCosta, B. F. "Arctic Exploration." *Journal of the American Geographical Society of New York* 12 (1880): 159–192.

———. "The Glacial Man in America." *Popular Science Monthly*, November 1880, 31–43.

De Haven, Edwin. "Official Report of the American Arctic Expedition." Appendix to *The United States Grinnell Expedition in Search of Sir John Franklin: A Personal Narrative*, by Elisha Kent Kane, 498. Philadelphia: Childs and Peterson, 1857.

De Long, Emma. *The Voyage of the* Jeannette: *The Ship and Ice Journals of George W. De Long, Lieutenant-Commander U.S.N. and Commander of the Polar Expedition of 1879–1881.* Boston: Houghton Mifflin, 1884.

Dickens, Charles. "The Lost Arctic Voyagers." *Littell's Living Age*, January 1855, 195–206.

Dickinson, Emily. *The Letters of Emily Dickinson.* Edited by Thomas H. Johnson, vol. 3. Cambridge: Harvard University Press, Belknap Press, 1958.

———. *The Poems of Emily Dickinson.* Edited by Thomas H. Johnson. Cambridge: Harvard University Press, Belknap Press, 1955.

Doten, Elizabeth. "A Song of the North." *Littell's Living Age*, January 1854, 193–194.

Doyle, A. Conan. *The Captain of the Pole Star.* Short Story Index Reprint Series. Freeport, NY: Books for Libraries, 1894.

———. "The Glamour of the Arctic." *McClure's Magazine*, March 1894, 391–400.

"Dr. Hayes' Exploration." *Littell's Living Age*, January 1862, 17–22.

"Dr. Kane's Arctic Expedition." *Congressional Globe* 24 (January 1855): 251.

"Dr. Kane's Arctic Operations." *Eclectic Magazine*, April 1857, 433–54.

"Dr. Kane's Expedition." *Littell's Living Age*, March 1856, 675–676.

Durand, Elias. "Floral Collections of Dr. Kane's Arctic Cruises." *Journal of the Academy of Natural Sciences* (1856).

Dwight, Theodore. "Arctic Meeting at Chickering Hall, November 21st, 1884." *Journal of the American Geographical Society of New York* 16 (1884): 338.

Dyche, Lewis Lindsay. "The Curious Race of Arctic Highlanders." *Cosmopolitan*, July 1896, 228–237.

Editorial. *Science* 4 (August 1884): 94.

Edselas, F. M. "Boston Half a Century Ago." *Catholic World*, March 1896, 733-746.

Elder, William. *Biography of Elisha Kent Kane*. Philadelphia: Childs and Peterson, 1858.

————. "Dr. Kane." *Littell's Living Age*, February 1856, 427-430.

"Electric Light for Polar Explorers." *Literary Digest*, 3 February 1906, 161.

Emerson, Ralph Waldo. *The Complete Works of Ralph Waldo Emerson*. Boston: Houghton Mifflin, 1903-1921.

"The Expedition in Search of Dr. Kane." *Littell's Living Age*, June 1855, 680.

"The Fate of Franklin and His Men." *Littell's Living Age*, December 1854, 529.

"The Fate of Sir John Franklin." *Littell's Living Age*, October 1854, 78-86.

Fiske, Daniel W. "The Progress of Marine Geography: Compiled by the General Secretary from Data Furnished by the Hydrographical Office, Washington." *Journal of the American Geographical and Statistical Society* 2 (1860): 1-12.

Foote, Henry S. "The Sir John Franklin Expedition." *Congressional Globe* 19 (May 1850): 884-887.

Ford, Frank Lewis. "The Heroes of the Icy North." *Munsey's Magazine*, December 1895, 286-296.

Free Public Libraries: Suggestions on Their Foundation and Administration. New York: American Social Science Association, 1871.

Galton, Francis. "Heredity Talent and Character." Pts. 1 and 2. *MacMillian's Magazine*, July 1865, 157-166; August 1865, 318-327.

Gilder, William H. "Arctic Lorelei Still Unconsenting." *Illustrated American*, October 1894, 453-456.

————. "An Expedition to the North Magnetic Pole." *McClure's Magazine*, July 1893, 159-162.

————. *Schwatka's Search: Sledging in the Arctic in Quest of the Franklin Records*. New York: Charles Scribner's Sons, 1881.

Gilman, Daniel. "Reception of the Greely Arctic Explorer, Lieutenant Greely, U.S.A." *Johns Hopkins University Circulars* 4 (March 1885): 54.

Greely, Adolphus. "Animals of the Arctic Region." *Chautauquan* 7 (May 1887): 468-470.

————. "Arctic Meeting at Chickering Hall, November 21st, 1884." *Journal of the American Geographical Society of New York* 16 (1884): 311-344.

————. "The Discoverers of the North Pole." *Munsey's Magazine*, November 1909, 290-296.

————. Introduction to *Discovery of the North Pole*, by J. Martin Miller. N.p.: J. T. Moss, 1909.

————. "Nansen's Polar Expedition." *National Geographic*, March 1896, 98-101.

————. "The Scope and Value of Arctic Explorations." *National Geographic*, January 1896, 32-39.

————. *Three Years of Arctic Service: An Account of the Lady Franklin Bay Expedition of 1881-84, and the Attainment of the Farthest North*. New York: C. Scribner's Sons, 1886.

————. "When I Stood Face to Face with Death." *Ladies' Home Journal*, October 1898, 2.

————. "The White North." *Pall Mall Gazette*. 7 November 1885, 1-2.

————. "Will They Reach the Pole?" *McClure's Magazine*, June 1894, 39-44.

"The Greely Arctic Expedition." *Medical News* 25 (September 1884): 265-266.

"The Greely Search." *Science* 3 (March 1884): 377-380.

Hall, Charles. *Arctic Researches and Life among the Esquimaux: Being the Narrative of an Expedition in Search of Sir John Franklin in the Years 1860, 1861, and 1862*. New York: Harper and Brothers, 1866.

Harris, R. A. "Some Indication of Land in the Vicinity of the North Pole." *National Geographic*, June 1904, 255-261.

Hawkes, Francis. *Uncle Philip's Conversation with the Children about the Whale Fishery and Polar Seas*, Boys and Girl's Library 26. New York: Harper and Brothers, 1836.

Hayes, Isaac I. "Arctic Meeting at Chickering Hall." *Journal of the American Geographical and Statistical Society* 12 (1880): 258-273.

———. "The Goblin of the Ice, or Christmas at the North Pole: A Legend of an Island in the Arctic Sea That Was Astray and Had Been Stolen." *Scribner's Monthly*, 1870 246–266.

———. "Observations on the Practicability of Reaching the North Pole." *Journal of Arts and Sciences* 76 (1858): 305–323.

———. *The Open Polar Sea: A Narrative of a Voyage of Discovery towards the North Pole, in the Schooner "United States."* New York: Hurd and Houghton, 1867.

———. "Physical Observations in the Arctic Seas." Smithsonian Contributions to Knowledge 15. Washington, DC: Smithsonian Institution Press, 1867.

———. *The Polar Exploring Expedition.* New York: Hurd and Houghton, 1867.

———. "Transactions of the Society for 1868." *Journal of the American Geographical and Statistical Society* 2, no. 2 (1868): xxxix–lvi.

———. "Transactions of the Society for 1870." *Journal of the American Geographical and Statistical Society* 2, no. 2 (1870): lxxi–lxxxi.

Hendrik, Hans. *Memoirs of Hans Hendrik.* London: Trubner, 1878.

Henry, Joseph. *A Scientist in American Life: Essays and Lectures of Joseph Henry.* Washington, DC: Smithsonian Institution Press, 1980.

Higbie, Edwin S. "The Aurora Borealis." *Littell's Living Age*, March 1850, 569.

Hill, John A. "The Polar Zone." *McClure's Magazine*, May 1898, 1–16.

"History, Biography, and Topography." *Methodist Quarterly Review* 27 (April 1867): 317–319.

House Committee on Naval Affairs. *Expedition to the Arctic Seas*, 45th Cong., 2nd sess., 1878, H. Rep. 96, 10–11 (testimony of Mr. Willis).

———. *Expedition to the Arctic Seas*, 46th Cong., 2nd sess., 1880, H. Rep. 453, 3 (testimony of Mr. Whitthorne).

Hubbard, Gardiner G. "Introductory Address." *National Geographic*, October 1888, 3–10.

"In Quest of the Pole." *Popular Science Monthly*, July 1873, 363–367.

Irving, Washington. *The Adventures of Captain Bonneville.* Edited by Robert A. Rees and Alan Sandy. Boston: Twayne, 1977.

———. *Astoria, or Anecdotes of an Enterprise Beyond the Rocky Mountains.* Philadelphia: Carey, Lea, and Blanchard, 1836.

"Is There an Open Arctic Sea?" *Catholic World*, October 1865, 137.

Johnson, John, Jr. "The Savagery of Boyhood." *Popular Science Monthly*, October 1887, 799.

Joint Committee. *Honors to Dr. Kane: Report of the Joint Committee Appointed to Receive the Remains and Conduct the Obsequies of the Late Elisha Kent Kane.* Philadelphia: James B. Chandler, 1857.

Kane, Elisha Kent. *Arctic Explorations: The Second Grinnell Expedition in Search of Sir John Franklin, 1853, 1854, 1855.* Philadelphia: Childs and Peterson, 1856.

———. *Astronomical Observations in the Arctic Seas.* Smithsonian Contributions to Knowledge 12. Washington, DC: Smithsonian Institution Press, 1860.

———. "Lecture on the Access to an Open Polar Sea in Connection with the Search after Sir John Franklin and His Companions, Read before the American Geographical and Statistical Society at Its Regular Monthly Meeting, by Dr. Kane, December 14, 1852." In *The United States Grinnell Expedition in Search of Sir John Franklin: A Personal Narrative*, 543–552. Philadelphia: Childs and Peterson, 1857.

———. *Magnetic Observations in the Arctic Seas.* Smithsonian Contributions to Knowledge 10. Washington, DC: Smithsonian Institution Press, 1858.

———. *The United States Grinnell Expedition in Search of Sir John Franklin: A Personal Narrative.* Philadelphia: Childs and Peterson, 1857.

Kane, Margaret Fox. *The Love-Life of Dr. Kane.* New York: Carleton, 1866.

Kane, Thomas L. *Alaska and the Polar Regions: Lecture of General Thomas L. Kane before the American Geographical Society in New York City, Thursday Evening, May 7, 1868.* New York: Journeymen Printers' Cooperative Association, 1868.

La Pérouse, Jean-François de Galaup de. *The Journal of Jean-François de Galaup de la Pérouse, 1785-1788.* London: Hakluyt Society, 1994-1995.

Longfellow, Henry Wadsworth. *The Complete Poetical Works of Longfellow.* Boston: Houghton Mifflin, 1922.

Lyon, G. F. "An Esquimaux Concert." *Littell's Living Age,* December 1854, 521.

Madden, E. F. "Symmes and His Theory." *Harper's New Monthly Magazine,* October 1882, 740-744.

"The Magnetic and Tidal Work of the Greely Arctic Expedition." *Science* 9 (March 1887): 215-217.

The Mariner's Chronicles: Containing Narratives of the Most Remarkable Disasters at Sea, Such as Shipwrecks, Storms, Fires, and Famines. New Haven: G. W. Gorton, 1834.

Markham, A. H. "The North Polar Problem." *North American Review* 162 (April 1896): 486-496.

Martin, Benjamin Nichols. *Choice Specimens of American Literature, and Literary Reader: Being Selections from the Chief American Writers.* New York: Sheldon, 1875.

Maury, Alfred. "Zoological Geography, or the Geographical Distribution of Animals." *Journal of the American Geographical and Statistical Society* 2 (1860): 68-91.

Maury, Matthew Fontaine. *The Physical Geography of the Sea.* New York: Harper and Brothers, 1855.

Maury, T. B. "Captain Hall's Arctic Expedition." *Galaxy* 11 (1871): 514-533.

———. "The Dumb Guides to the Pole." *Putnam's Monthly Magazine,* December 1869, 727-741.

———. "The Eastern Portal to the Pole." *Putnam's Monthly Magazine,* April 1870, 437-445.

———. "The Gateways to the Pole." *Putnam's Monthly Magazine,* November 1869, 521-537.

Mayo, Earl. "The Ice-Breaker 'Ermack.'" *McClure's Magazine,* April 1900, 537-544.

McGinley, William. *Reception of Lieut. A. W. Greely, U.S.A., and His Comrades, and of the Arctic Relief Expedition, at Portsmouth, N. H., on August 1 and 4, 1884.* Washington, DC: Government Printing Office, 1884.

McLane, Robert. "Franklin Expedition." *Congressional Globe* 19 (April 1850): 835.

Meech, L. W. *On the Relative Intensity of the Heat and Light of the Sun upon Different Latitudes of the Earth.* Smithsonian Contributions to Knowledge 9. Washington, DC: Smithsonian Institution Press, 1857.

Melville, George. *In the Lena Delta.* Boston: Houghton Mifflin. 1885.

Mill, Hugh Robert. "The Race to the North Pole: The Expeditions of Nansen and Jackson." *McClure's Magazine,* July 1893, 146-155.

———. "Unknown Parts of the World," *McClure's Magazine,* November 1894, 540-550.

Miller, Jacob. "The Sir John Franklin Expedition." *Congressional Globe* 19 (May 1850): 884-886.

Miller, J. Martin. *Discovery of the North Pole.* N.p.: J. T. Moss, 1909.

"Miscellaneous." *Littell's Living Age,* March 1855, 744.

"A Modern Viking." *National Geographic,* January 1906, 38-39.

Moffett, Cleveland. "Lieutenant Peary's Expedition." *McClure's Magazine,* July 1893, 156-158.

Morton, William. *Dr. Kane's Arctic Voyage: Explanatory of a Pictorial Illustration of the Second Grinnell Expedition.* Boston: Barton and Son, 1857.

"A Motor for Polar Snow-Fields." *Literary Digest,* 21 May 1910, 1026.

"Nordenskiold's Arctic Investigations." *Science* 5 (May 1885): 430-432.

Northend, Charles. *The Teacher's Assistant, or Hints and Methods in School Discipline and Instruction: Being a Series of Familiar Letters to One Entering upon the Teacher's Work.* Chicago: G. C. W Sherwood, 1865.

Nourse, J. E. *Narrative of the Second Arctic Expedition.* Washington, DC: Government Printing Office, 1879.

"Nourse's American Exploration in the Ice-Zones." *Science* 3 (June 1884): 766.

Ogden, Herbert G. "Report—Geography of the Land." *National Geographic,* April 1889, 125–135.

Olmsted, Denison. *On the Recent Secular Period of the Aurora Borealis.* Smithsonian Contributions to Knowledge 7. Washington, DC: Smithsonian Institution Press, 1856.

"An Open Polar Ocean." *Eclectic Magazine of Foreign Literature, Science, and Art,* January 1872, 113–115.

"The Open Polar Sea." *Littell's Living Age,* April 1867, 59–62.

"The Open Polar Sea." *Atlantic Monthly,* April 1867, 511–512.

P., M. "The Capriciousness of Public Notice." *Earnest Worker,* March 1887, n.p.

Parry, W. E. *Three Voyages for the Discovery of a Northwest Passage from the Atlantic to the Pacific, and Narrative of an Attempt to Reach the North Pole.* Harper's Family Library 108. New York: Harper and Brothers, 1840.

Pavy, Lilly May. "Dr. Pavy and the Polar Expedition." *North American Review* 142 (March 1886): 258–269.

Pearson, Karl. *The Life, Letters, and Labours of Francis Galton.* Cambridge: Cambridge University Press, 1914.

Peary, Josephine. *The Snow Baby: A True Story with True Pictures.* New York: Frederick Stokes, 1901.

Peary, Robert. "Gold Medal of the Paris Geographical Society Presented to Commander Peary." *Journal of the American Geographical Society of New York* 36 (1904): 693–694.

———. *Nearest the Pole: A Narrative of the Polar Expedition of the Peary Arctic Club in the S. S. Roosevelt, 1905–1906.* New York: Doubleday, Page, 1907.

———. *The North Pole: Its Discovery in 1909 under the Auspices of the Peary Arctic Club.* New York: Frederick Stokes, 1910.

———. *Northward over the "Great Ice": A Narrative of Life and Work along the Shores and upon the Interior Ice-Cap of Northern Greenland in the Years 1886 and 1891–1897,* vol. 1. New York: Frederick Stokes, 1898.

———. "Peary and the North Pole." *National Geographic,* October 1903, 379–381.

———. "Peary on the North Pole." *National Geographic,* January 1903, 29.

———. "Peary's Work in 1901–1902." *National Geographic,* October 1902, 384–386.

"Peary's New Arctic Ship 'Roosevelt.'" *Literary Digest,* 27 May 1905, 780–781.

"Peary's 'Proofs.'" *Literary Digest,* 23 October 1909, 659–661.

Peck, Harry Thurston. "The Disputes of Great Discoverers." *Munsey's Magazine,* January 1910, 476.

Petermann, Augustus. "On the Geography of the Arctic Regions." *Littell's Living Age,* January 1856, 205–207.

Poe, Edgar Allen. *Narrative of Arthur Gordon Pym.* London: J. Cunningham, 1841.

"Polar Climate in Time the Major Factor in the Evolution of Plants and Animals." *Journal of the American Geographical Society of New York* 36(1904): 142–145.

"The Polar Cryptogram." *Literary Digest,* 30 October 1909, 738.

"Polar Discovery and Controversy." *Literary Digest,* 18 September 1909, 417–419.

"Polar Perpetrations." *Literary Digest,* 9 October 1909, 603–606.

Prentiss, Henry Mellen. *The Great Polar Current.* Cambridge, MA: Riverside, 1897.

Preston, William Ballard. "The Arctic Expedition at Sea—Lt. De Haven's Instructions." *Littell's Living Age,* July 1850, 45–46.

Proceedings of the Academy of Natural Sciences of Philadelphia. Philadelphia: Merrihew and Thompson, 1852.

Proceedings of the Academy of Natural Sciences of Philadelphia. Philadelphia: Merrihew and Thompson, 1853.

Proceedings of the Academy of Natural Sciences of Philadelphia. Philadelphia: Merrihew and Thompson, 1857.

Proceedings of the Academy of Natural Sciences of Philadelphia. Philadelphia: Merrihew and Thompson, 1858.

Proceedings of the Academy of Natural Sciences of Philadelphia. Philadelphia: Merrihew and Thompson, 1859.

Proceedings of the Academy of Natural Sciences of Philadelphia. Philadelphia: Merrihew and Thompson, 1861.

Proceedings of the Academy of Natural Sciences of Philadelphia. Philadelphia: Merrihew and Thompson, 1862.

Proceedings of the Academy of Natural Sciences of Philadelphia. Philadelphia: Merrihew and Thompson, 1863.

"A Question of the Poles." *American Review of Reviews* (December 1909): 736–737.

Rae, John. *John Rae's Arctic Correspondence, 1844–55*. Edited by E. E. Rich. Publications of the Hudson's Bay Record Society 16. London: Hudson's Bay Record Society, 1953.

———. "The Lost Arctic Voyagers." *Littell's Living Age*, February 1855, 451–456.

———. "Miscellaneous." *Littell's Living Age*, April 1855, 184.

———. "Remains of Sir John Franklin and His Party." *Littell's Living Age*, November 1854, 311–312.

"The Recent Arctic Expeditions." *Littell's Living Age*, April 1850, 18–36.

"Recent Polar Explorations." *Popular Science Monthly*, June 1875, 320–332.

Reclus, Elisée. *North America*. Vol. 1, *The Earth and Its Inhabitants*. Edited by A. H. Keane. New York: D. Appleton, 1893.

Reid, Harry Fielding. "How Could an Explorer Find the Pole?" *Popular Science Monthly*, January 1910, 89–97.

"Relief to Dr. Kane." *Congressional Globe* 24 (January 1855): 229.

"Results of the Arctic Search." *North American Review* 84 (January 1857): 96.

Rink, H. J. "The Polar Question." *Proceedings of the United States Naval Institute* 11 (1885): 633–699.

Robeson, George M. *Instructions for the Expedition toward the North Pole*. Washington, DC: Government Printing Office, 1871.

Roosevelt, Theodore. "The American Boy." In *The Strenuous Life: Essays and Addresses*, 2. New York: Century, 1900.

Ross, John. *Narrative of a Second Voyage in Search of a North-West Passage, and of a Residence in the Arctic Regions during the Years 1829, 1830, 1831, 1832, 1833*. Philadelphia: Carey and A. Hart, 1835.

Rosse, Irving C. "The First Landing on Wrangel Island, with Some Remarks on the Northern Inhabitants." *Journal of the American Geographical Society of New York* 15 (1883): 163–215.

St. John, Percy B. *The Arctic Crusoe: A Tale of the Polar Sea*. Boston: Lee and Shepard, 1890.

Sargent, Epes. *Arctic Adventure by Land and Sea*. Boston: Phillips, Sampson, 1857.

Savage, John. "Franklin Expedition." *Congressional Globe* 19 (April 1850): 832.

Savage, M. J. *Our Heroic Age*, vol. 6. Boston: George H. Ellis, 1884.

Schley, W. S. *The Rescue of Greely*. New York: Charles Scribner's Sons, 1885.

Schwatka, Frederick. "The Alaska Military Reconnaissance of 1883." *Science* 3 (February 1884): 220–227.

———. "An Arctic Vessel and Her Equipment." *Science* 3 (April 1884): 505–511.

———. "Coming Polar Expeditions." *North American Review* 148 (1889): 154.

————. "Exploration of the Yukon River in 1883." *Journal of the American Geographical Society of New York* 16 (1884): 345–382.

————. "Icebergs and Ice-floes." *Science* 3 (May 1884): 535–538.

————. "Letter of Lieut. Frederick Schwatka, U.S. Army, Commanding Franklin Search Party." *Journal of the American Geographical Society of New York* 12 (1880): 104–107.

————. "Wintering in the Arctic." *Science* 3 (May 1884): 566–571.

"Science and Polar 'Dashes.'" *Literary Digest*, 9 October 1909, 573.

Seaborne, Adam. *Symzonia: A Voyage of Discovery*. New York: J. Seymour, 1820.

"Search for Dr. Kane." *Congressional Globe* 24 (December 1854): 82.

"The Search for Sir John Franklin." *Littell's Living Age*, February 1850, 279–281.

"Second American Expedition in Search of Sir John Franklin." *Littell's Living Age*, April 1853, 35–36.

Seward, William H. "The Sir John Franklin Expedition." *Congressional Globe* 19 (May 1850): 885–886.

Shelley, Mary. *Frankenstein; or, The Modern Prometheus*. New York: Doubleday, 1999.

Shields, Charles W. *Funeral Eulogy at the Obsequies of Dr. E. K. Kane*. Philadelphia: Parry and McMillian, 1857.

Sigourney, Lydia. "The King of Icebergs." *Lady's Book*, July 1841, 43.

"Sir John Franklin and the Arctic Regions." *North American Review* 71 (July 1850): 168–185.

"The Sir John Franklin Expedition." *Congressional Globe* 19 (May 1850): 884–891.

Smith, Francis H. *My Experience; or, Footprints of a Presbyterian to Spiritualism*. Baltimore: n.p., 1860.

"Smith Sound, and Its Exploration." *Science* 3 (May 1884): 622–623.

"Snow's Voyage on the Prince Albert." *Littell's Living Age*, March 1851, 421.

Sonntag, August. *Professor Sonntag's Thrilling Narrative of the Grinnell Exploring Expedition to the Arctic Ocean, in the Years 1853, 1854, and 1855, in Search of Sir John Franklin, under the Command of Dr. E. K. Kane*. Philadelphia: J. T. Lloyd, 1857.

Stadling, Jonas. "Andrée's Flight into the Unknown." *Century Magazine*, November 1897, 81–88.

Stimpson, William. *Synopsis of the Marine Invertebrata Collected by the Late Arctic Expedition, under Dr. I. I. Hayes*. Philadelphia: n.p., 1863.

Stokes, Frederick Wilbert. "An Arctic Studio." *Century Magazine*, July 1896, 408–414.

————. "Color at the Far North." *Century Magazine*, September 1894, 722–724.

Thoreau, Henry David. *Life without Principle*. Norwood, PA: Norwood Editions, 1978.

"Tires for Arctic Use." *Literary Digest*, 25 January 1908, 120.

"To the North Pole by Automobile." *Literary Digest*, 14 June 1902, 805.

"Transactions of the Society for 1868." *Journal of the American Geographical and Statistical Society* 2 (1868): xxxix–lxxxi.

"The United States Grinnell Expedition in Search of Sir John Franklin." *North American Review* 80 (April 1855): 307–342.

U.S. Congress. *Congressional Record*. 47th Cong., 1st sess., 1881. Washington, DC: Government Printing Office, 1203.

Wagenen, Theodore F. Van. "Commercial Possibilities of the North Polar Region." *Conservative Review*, June 1900, 392.

Walker, D. "Captain Hall's Arctic Expedition." *Overland Monthly*, September 1871, 201–208.

"Walter Wellman's Expedition to the North Pole." *National Geographic*, April 1906, 205–207.

Weildon, W. W. "Remarks on the Supposed Open Sea in the Arctic Regions." *Proceedings of the American Association for the Advancement of Science* 14 (1860): 166–174.

Wellman, Walter. *The Aerial Age: A Thousand Miles by Airship over the Atlantic Ocean*. New York: A. R. Keller. 1911.

————. "An Arctic Day and Night." *McClure's Magazine*, April 1900, 555–563.

————. "By Airship to the North Pole." *McClure's Magazine*, June 1907, 189-200.

————. "Long-Distance Balloon Racing." *McClure's Magazine*, July 1901, 203-214.

————. "On the Way to the Pole: The Wellman Polar Expedition." *Century Magazine*, February 1899, 530-537.

————. "The Polar Airship." *National Geographic*, April 1906, 208-228.

————. "The Race for the North Pole." *McClure's Magazine*, February 1900, 318-328.

————. "Sledging toward the Pole." *McClure's Magazine*, March 1900, 405-414.

"Wellman Polar Expedition." *National Geographic*, August 1898, 373-375.

"The Wellman Polar Expedition." *National Geographic*, December 1906, 712.

Wells, H. G. *The Time Machine: An Invention*. New York: H. Holt, 1895.

————. *The War of the Worlds*. New York: Harper and Brothers, 1898.

Whitman, Walt. *Leaves of Grass*. Brooklyn: n.p., 1855.

Whitney, Harry. "Harry Whitney's Arctic Hunt." *Literary Digest*, 23 October 1909, 692-693.

Wilkes, Charles. *Narrative of the U.S. Exploring Expedition during the Years 1838, 1839, 1848, 1841, 1842*, vol. 1. Philadelphia: Lea and Blanchard, 1845.

Wilkinson, W. C. "My Open Polar Sea," *Scribner's Monthly*, June 1875, 194.

————. "The Northern Lights," *Scribner's Monthly*, January 1870, 310.

Williams, Edwin. *Narrative of the Recent Voyage of Captain Ross to the Arctic Regions in the Years 1829, 1830, 1831, 1832, 1833*. New York: Wiley and Long, 1835.

SECONDARY SOURCES

Adas, Michael. *Machines as the Measure of Men: Science, Technology, and Ideologies of Western Dominance*. Edited by George Fredrickson and Theda Skocpol. Cornell Studies in Comparative History. Ithaca: Cornell University Press, 1989.

American National Biography. Edited by John Garraty and Mark C. Carnes. New York: Oxford University Press, 1999.

Anderson, J. R. L. *The Ulysses Factor: The Exploring Instinct in Man*. London: Hodder and Stoughton, 1970.

Baatz, Simon. *Knowledge, Culture, and Science in the Metropolis: The New York Academy of Science, 1817-1870*. New York: New York Academy of Science, 1990.

Barr, William. *The Expeditions of the First International Polar Year, 1882-83*. Calgary: Arctic Institute of North America, University of Calgary, 1985.

Baur, John I. H. "A Romantic Impressionist: James Hamilton." *Bulletin of the Brooklyn Museum* 12 (Spring 1951): 1-8.

Bederman, Gail. *Manliness and Civilization: A Cultural History of Gender and Race in the United States, 1880-1917*. Edited by Catherine R. Stimpson. Women in Culture and Society. Chicago: University of Chicago Press, 1995.

Bell, Michael Davitt. "Conditions of Literary Vocation." In *The Cambridge History of American Literature: Prose Writing, 1820-1865*, edited by Sacvan Bercovitch, 9-123. New York: Cambridge University Press, 1995.

Berkhofer, Robert F. "White Conceptions of Indians." In *History of Indian-White Relations*, edited by Wilcomb E. Washburn, 522-547. Handbook of North American Indians 4. Washington: Smithsonian Institution, 1988.

————. *The White Man's Indian: Images of the American Indian from Columbus to the Present*. New York: Vintage Books, 1979.

Berton, Pierre. *The Arctic Grail: The Quest For the Northwest Passage and the North Pole, 1818-1909*. New York: Viking, 1988.

Blanchard, Mary Warner. *Oscar Wilde's America: Counterculture in the Gilded Age*. New Haven: Yale University Press, 1998.

Bloom, Lisa. *Gender on Ice: American Ideologies of Polar Expeditions.* Minneapolis: University of Minneapolis Press, 1993.

Brown, Kathleen M. "Brave New Worlds: Women's and Gender History." *William and Mary Quarterly* 50 (April 1993): 311-328.

Browne, Janet. *Charles Darwin: A Biography.* New York: Knopf, 1995-2002.

Bruce, Robert V. *The Launching of American Science.* New York: Knopf, 1987.

Bryce, Robert M. *Cook and Peary: The Polar Controversy, Resolved.* Mechanicsburg, PA: Stackpole Books, 1997.

Burnett, D. Graham. *Masters of All They Surveyed: Exploration, Geography, and a British El Dorado.* Chicago: University of Chicago Press, 2002.

Capelotti, P. J. *By Airship to the North Pole: An Archaeology of Human Exploration.* New Brunswick, NJ: Rutgers University Press, 1999.

Caputi, Jane. "The Metaphors of Radiation, or Why a Beautiful Woman Is Like a Nuclear Power Plant." *Women's Studies International Forum* 14 (1991): 423-442.

Carnes, Mark C. *Secret Ritual and Manhood in Victorian America.* New Haven: Yale University Press, 1989.

Carnes, Mark C., and Clyde Griffen, eds. *Meanings for Manhood.* Chicago: University of Chicago Press, 1990.

Carr, Gerald L. *Frederick Edwin Church: The Icebergs.* Dallas: Dallas Museum of Fine Arts, 1980.

Caswell, John Edwards. *Arctic Frontiers: United States Explorations in the Far North.* Norman: University of Oklahoma Press, 1956.

Chapin, David. "'Science Weeps, Humanity Weeps, the World Weeps': America Mourns Elisha Kent Kane." *Pennsylvania Magazine of History and Biography* 123 (1999): 275-301.

Cochran, Thomas C. *200 Years of American Business.* New York: Basic Books, 1977.

Cockburn, Cynthia. *Brothers: Male Dominance and Technical Change.* London: Pluto, 1983.

Cole, Douglas. *Franz Boas: The Early Years, 1858-1906.* Vancouver: Douglas and McIntyre, 1999.

Collis, Christy. "The Voyage of the Episteme: Narrating the North." *Essays on Canadian Writing* 59 (Fall 1996): 26-45.

Cooter, Roger, and Stephen Pumfrey. "Separate Spheres and Public Places: Reflections on the History of Science Popularization and Science in Popular Culture." *History of Science* 32 (September 1994): 237-267.

Corner, George W. *Dr. Kane of the Arctic Seas.* Philadelphia: Temple University Press, 1972.

Cowan, Ruth Schwartz. *A Social History of American Technology.* New York: Oxford University Press, 1997.

Crouch, Tom D. *The Eagle Aloft: Two Centuries of the Balloon in America.* Washington, DC: Smithsonian Institution Press, 1983.

Damon-Moore, Helen. *Magazines for the Millions: Gender and Commerce in the Ladies' Home Journal and the Saturday Evening Post, 1880-1910.* Albany: State University of New York Press, 1994.

Daniels, George H. *Science in American Society: A Social History.* New York: Knopf, 1971.

Davis, Lance E., Robert E. Gallman, and Karin Gleiter. *In Pursuit of Leviathan: Technology, Institutions, Productivity, and Profits in American Whaling, 1816-1906.* Chicago: University of Chicago Press, 1997.

Dawson, Graham. *Soldier Heroes: British Adventure, Empire, and the Imagining of Masculinities.* London: Routledge, 1994.

Dawson, Michael. Review of "Wild Things: Nature, Culture, and Tourism in Ontario, 1790-1914." H-Canada, N-Net Reviews, January 1997. Http://www.h-net.org/reviews/showrev.cgi?path=12175862319888.

Derounian-Stodola, Katherine Zabelle. *The Indian Captivity Narrative, 1550-1900.* New York: Twayne, 1993.

DeVorkin, David H. *Race to the Stratosphere: Manned Scientific Ballooning in America*. New York: Springer-Verlag, 1989.

Driver, Felix. *Geography Militant: Cultures of Exploration and Empire*. Oxford: Blackwell, 2001.

Dupree, A. Hunter. *Science in the Federal Government*. Cambridge: Harvard University Press, Belknap Press, 1957.

Fabian, Ann. "The Ragged Edge of History: Intellectuals and the American West." *Reviews in American History* 26 (1998): 575–580.

Fleming, Fergus. *Barrow's Boys*. New York: Atlantic Monthly Press, 2000.

Fleming, James. *Meteorology in America, 1800–1870*. Baltimore: Johns Hopkins University Press, 1990.

Godwin, Joscelyn. *Arktos: The Polar Myth in Science, Symbolism, and Nazi Survival*. London: Thames and Hudson, 1993.

Goetzmann, William H. *Army Exploration of the American West, 1803–1863*. New Haven: Yale University Press, 1959.

———. *Exploration and Empire: The Explorer and the Scientist in the Winning of the American West*. New York: Norton, 1966.

Guttridge, Leonard F. *Ghosts of Cape Sabine: The Harrowing True Story of the Greely Expedition*. New York: Berkley Books, 2000.

———. *Icebound: The Jeannette Expedition's Quest for the North Pole*. Shrewsbury, UK: Airlife, 1987.

Harper, Kenn. *Give Me My Father's Body: The Life of Minik, the New York Eskimo*. Frobisher Bay, NWT: Blacklead Books, 1986.

Hawes, Joseph M. "The Signal Corps and Its Weather Service, 1870–1890." *Military Affairs* 30 (Summer 1966): 69.

Hellman, Geoffrey. *Bankers, Bones, and Beetles: The First Century of the American Museum of Natural History*. Garden City, NY: Natural History Press, 1968.

Herbert, Wally. *The Noose of Laurels: Robert Peary and the Race to the North Pole*. New York: Athenaeum, 1989.

Herman, Daniel J. "The Other Daniel Boone: The Nascence of a Middle-Class Hunter Hero, 1784–1860." *Journal of the Early Republic* 18 (Fall 1998): 429–457.

Herzig, Rebecca. "In the Name of Truth: Sacrificial Ideals and American Science, 1870–1930." Ph.D. diss., Massachusetts Institute of Technology, 1998.

Hevly, Bruce. "The Heroic Science of Glacier Motion." *Osiris* 11 (1996): 66–86.

Hindle, Brooke. *The Pursuit of Science in Revolutionary America, 1735–1789*. Chapel Hill: University of North Carolina Press, 1956.

Hobbs, William Herbert. *Peary*. New York: Macmillan, 1936.

Holland, Clive. *Arctic Exploration and Development, c. 500 B.C. to 1915: An Encyclopedia*. New York: Garland, 1994.

———. *Farthest North: The Quest for the North Pole*. New York: Carroll and Graf, 1994.

Huhndorf, Shari. "Going Native: Figuring the Indian in Modern American Culture." Ph.D. diss., New York University, 1996.

Huntford, Roland. *The Last Place on Earth*. New York: Athenaeum, 1983.

———. *Nansen: The Explorer as Hero*. London: Duckworth, 1997.

Huntington, David C. *The Landscapes of Frederic Edwin Church: Vision of an American Era*. New York: George Braziller, 1966.

James, P. E., and G. J. Martin. *The Association of American Geographers: The First Seventy-Five Years*. [Washington, DC: The Assocation], 1978.

Joyce, Barry Alan. "'As the Wolf from the Dog': American Overseas Exploration and the Compartmentalization of Mankind, 1838–1859." Ph.D. diss., University of California at Riverside, 1995.

———. "Elisha Kent Kane and the Eskimo of Etah." In *Surveying the Record: North American Scientific Exploration to 1930*, edited by Edward C. Carter, 103–117. Memoirs of the American Philosophical Society. Philadelphia: American Philosophical Society, 1999.

Kane, Anne. "Reconstructing Culture in Historical Explanation: Narratives as Cultural Structure and Practice." *History and Theory* 39 (October 2000): 311–330.

Kasson, John F. *Civilizing the Machine: Technology and Republican Values in America, 1776–1900.* New York: Grossman, 1976.

Keeney, Elizabeth Barnaby. *The Botanizers: Amateur Scientists in Nineteenth-Century America.* Chapel Hill: University of North Carolina Press, 1992.

Kennedy, John Michael. "Philanthropy and Science in New York City: The American Museum of Natural History, 1868–1968." Ph.D. diss., Yale University, 1968.

Kimmel, Michael. *Manhood in America: A Cultural History.* New York: Free Press, 1996.

Kitcher, Philip. "Persuasion." In *Persuading Science: The Art of Scientific Rhetoric*, edited by Marcello Pera and William R. Shea, 3–27. Canton, MA: Science History Publications, 1991.

Kohlstedt, Sally Gregory. "Creating a Forum for Science: AAAS in the Nineteenth Century." In *The Establishment of Science in America: 150 Years of the American Association for the Advancement of Science*, edited by Michael M. Sokal, Sally Kohlstedt, and Bruce V. Lewenstein, 7–49. New Brunswick: Rutgers University Press, 1999.

LaFollette, Marcel C. *Making Science Our Own: Public Images of Science, 1910–1955.* Chicago: University of Chicago Press, 1990.

Lears, T. J. Jackson. *No Place of Grace: Antimodernism and the Transformation of American Culture.* New York: Pantheon, 1981.

Lenz, William E. "Narratives of Exploration, Sea Fiction, Mariners' Chronicles, and the Rise of American Nationalism: 'To Cast Anchor on That Point Where All Meridians Terminate.'" *American Studies* 32 (Fall 1991): 41–62.

Lerman, Nina E., Arwen Palmer Mohun, and Ruth Oldenziel. "The Shoulders We Stand on and the View from Here: Historiography and Directions for Research." *Technology and Culture* 38 (January 1997): 9–30.

Levere, Trevor H. *Science and the Canadian Arctic: A Century of Exploration, 1818–1918.* Cambridge: University of Cambridge Press, 1993.

———. "Vilhjalmur Stefansson, the Continental Shelf, and a New Arctic Continent." *British Journal of the History of Science* 21 (1988): 233–247.

Levine, Lawrence W. *The Unpredictable Past: Explorations in American Cultural History.* Oxford: Oxford University Press, 1993.

Loomis, Chauncey C. *Weird and Tragic Shores: The Story of Charles Francis Hall, Explorer.* New York: Knopf, 1971.

Lyotard, Jean-François. *The Postmodern Condition: A Report on Knowledge.* Minneapolis: University of Minneapolis Press, 1993.

Mabley, Edward. *The Motor Balloon "America."* Brattleboro: Stephen Greene, 1969.

Machor, James L. *Pastoral Cities: Urban Ideals and the Symbolic Landscape of America.* Madison: University of Wisconsin Press, 1987.

Madsen, Deborah L. *American Exceptionalism.* Jackson: University Press of Mississippi, 1998.

Manthorne, Katherine E. "Legible Landscapes: Text and Image in the Expeditionary Art of Frederic Church." In *Surveying the Record: North American Scientific Exploration to 1930*, edited by Edward C. Carter, 133–145. Philadelphia: American Philosophical Society, 1999.)

———. *Tropical Renaissance: North American Artists Exploring Latin America, 1839–1879.* Washington, DC: Smithsonian Institution Press, 1989.

Marsh, Margaret. "Suburban Men and Masculine Domesticity, 1870–1915." *American Quarterly* 40 (June 1988): 165–186.

McCannon, John. *Red Arctic: Polar Exploration and the Myth of the North in the Soviet Union, 1932–1939.* New York: Oxford University Press, 1998.

Meining, Donald. *The Shaping of America: A Geographical Perspective on 500 Years of History*. New Haven: Yale University Press, 1986.

Miller, Angela. *The Empire of the Eye: Landscape Representation and American Cultural Politics, 1825–1875*. Ithaca: Cornell University Press, 1993.

Mirsky, Jeanette. *Elisha Kent Kane and the Seafaring Frontier*. Boston: Little, Brown, 1954.

————. *To the Arctic: The Story of Northern Exploration from the Earliest Times to the Present*. Chicago: University of Chicago Press, 1948.

Morse, David. *American Romanticism: From Cooper to Hawthorne*, vol. 1. Totowa, NJ: Barnes and Noble Books, 1987.

Mosse, George L. *The Image of Man: The Creation of Modern Masculinity*. New York: Oxford University Press, 1996.

Mott, Frank Luther. *American Journalism: A History, 1690–1960*, 3d ed. New York: Macmillan, 1962.

————. *A History of American Magazines*. Vol. 2, *A History of American Magazines, 1850–1865*. Cambridge: Harvard University Press, 1938.

Nash, Roderick. "The American Cult of the Primitive." *American Quarterly* 18 (1966): 517–537.

National Aeronautics and Space Administration. "The Vision for Space Exploration, February 2004," http://www.nasa.gov/pdf/55583main_vision_space_exploration2.pdf.

Neatby, Leslie. *In Quest of the Northwest Passage*. Toronto: Longmans and Green, 1958.

Novak, Barbara. *Nature and Culture: American Landscape and Painting, 1825–1875*. New York: Oxford University Press, 1980.

Nye, David E. *American Technological Sublime*. Cambridge: MIT Press, 1994.

Ohmann, Richard. *Selling Culture: Magazines, Markets, and Class at the Turn of the Century*. London: Verso, 1996.

Oldenziel, Ruth. *Making Technology Masculine: Men, Women, and Modern Machines in America, 1870–1945*. Amsterdam: Amsterdam University Press, 1999.

Pauly, Philip J. "The World and All That Is in It: The National Geographic Society, 1888–1918." *American Quarterly* 31 (1979): 517–532.

Pegg, Barry. "Nature and Nation in Popular Scientific Narratives of Polar Exploration." In *The Literature of Science: Perspectives on Popular Science Writing*, edited by Murdo William McRae, 213–229. Athens: University of Georgia Press, 1993.

Pera, Marcello, and William R. Shea. Preface to *Persuading Science: The Art of Scientific Rhetoric*, edited by Marcello Pera and William R. Shea, vii–x. Canton, MA: Science History Publications, 1991.

Poole, Deborah. "Landscape and the Imperial Subject: U.S. Images of the Andes, 1859–1930." In *Close Encounters of Empire: Writing the Cultural History of U.S.–Latin American Relations*, edited by Catherine C. LeGrand, Gilbert M. Joseph, and Ricardo D. Salvatore, 107–138. Durham Duke University Press, 1998.

Potter, Russell A. "The Sublime Yet Awful Grandeur: The Arctic Panoramas of Elisha Kent Kane." *Polar Record* 35 (July 1999): 194.

Pratt, Mary Louise. *Imperial Eyes: Travel Writing and Transculturation*. London: Routledge, 1992.

Preston, Douglas J. *Dinosaurs in the Attic: An Excursion into the American Museum of Natural History*. New York: St. Martin's, 1986.

Raines, Rebecca Robbins. *Getting the Message Through: A Branch History of the U.S. Army Signal Corps*. Washington, DC: Center for Military History, U.S. Army, 1996.

Riffenburgh, Beau. *The Myth of the Explorer: The Press, Sensationalism, and Geographical Discovery*. London: Belhaven, 1993.

Rink, Oliver A. *Holland on the Hudson: An Economic and Social History of Dutch New York*. Ithaca: Cornell University Press, 1986.

Roberts, Brian. *American Alchemy: The California Gold Rush and Middle-Class Culture*. Cultural Studies of the United States. Chapel Hill: University of North Carolina Press, 2000.

Rogin, Michael Paul. *Fathers and Children: Andrew Jackson and the Subjugation of the American Indian*. New York: Knopf, 1975.

Ross, W. Gillies. "Clairvoyants and Mediums Search for Franklin." *Polar Record* 39 (2003): 1-18.

————. "Nineteenth-Century Exploration of the Arctic." In *A Continent Comprehended*, edited by John Logan Allen, 244-331. Lincoln: University of Nebraska Press, 1997.

Rossiter, Margaret W. *Women Scientists in America: Struggles and Strategies to 1940*. Baltimore: Johns Hopkins University Press, 1982.

Rotundo, E. Anthony. *American Manhood: Transformations in Masculinity from the Revolutionary to the Modern Era*. New York: Basic Books, 1993.

————. "Body and Soul: Changing Ideals of American Middle-Class Manhood, 1770-1920." *Journal of Social History* 16 (1983): 32.

Rozwadowski, Helen M. *Fathoming the Ocean: The Discovery and Exploration of the Deep Sea*. Cambridge: Harvard University Press, Belknap Press, 2005.

Rubin, Joan Shelley. *The Making of Middlebrow Culture*. Chapel Hill: University of North Carolina Press, 1992.

Savours, Ann. *The Search for the North West Passage*. New York: St. Martin's, 1999.

Schatzberg, Eric. *Wings of Wood, Wings of Metal: Culture and Technical Choice in American Airplane Materials, 1914-1945*. Princeton: Princeton University Press, 1999.

Scholnick, Robert J. "Permeable Boundaries: Literature and Science in America." In *American Literature and Science*, edited by Robert J. Scholnick, 1-17. Lexington: University Press of Kentucky, 1992.

Schulten, Susan. *The Geographical Imagination in America, 1880-1950*. Chicago: University of Chicago Press, 2001.

Scott, Donald M. "The Popular Lecture and the Creation of a Public in Mid-Nineteenth-Century America." *Journal of American History* 66 (March 1980): 791-809.

Scott, Joan W. "Gender: A Useful Category of Historical Analysis." *Journal of American History* 91 (1986): 1053-1075.

Shapin, Steven. *A Social History of Truth: Civility and Science in Seventeenth-Century England*. Chicago: University of Chicago Press, 1994.

Sicherman, Barbara. "The Uses of a Diagnosis: Doctors, Patients, and Neurasthenia," *Journal of the History of Medicine and Allied Sciences* 32 (1977): 33-54.

Slotten, Hugh. *Patronage, Practice, and the Culture of American Science: Alexander Dallas Bache and the U.S. Coast Survey*. Cambridge: Cambridge University Press, 1994.

Spencer, John. "'We are not entirely dealing with the past': America Remembers Lewis and Clark." In *Lewis and Clark: Legacies, Memories, and New Perspectives*, edited by Kris Fresonke and Mark Spence, 159-183. Berkeley: University of California Press, 2004.

Spengemann, William C. *The Adventurous Muse: The Poetics of American Fiction, 1789-1900*. New Haven: Yale University Press, 1977.

Sterling, Keir B. *Last of the Naturalists: The Career of C. Hart Merriam*. New York: Arno, 1974.

Stocking, George W. *The Shaping of American Anthropology, 1883-1911: A Franz Boas Reader*. New York: Basic Books, 1974.

Sundquist, Eric J. "The Literature of Expansion and Race." In *The Cambridge History of American Literature*, ed. Sacvan Bercovitch, 178-80. New York: Cambridge University Press, 1995.

Taylor, Bryan C. "Register of the Repressed: Women's Voice and the Body in the Nuclear Weapons Organization." *Quarterly Journal of Speech* 97 (1993): 267-285.

Tebbel, John. *A History of Book Publishing in the United States: The Creation of an Industry, 1630-1865*. New York: R. R. Bowker, 1972.

Terrall, Mary. "Heroic Narratives of Quest and Discovery." *Configurations* 6 (1998): 223-242.

Thompson, Peter. "No Chance in Nature: Cannibalism as a Solution to Maritime Famine, c. 1750-1800." In *American Bodies: Cultural Histories of the Physique*, edited by Tim Armstrong, 32-44. New York: New York University Press, 1996.

Todd, A. L. *Abandoned*. New York: McGraw-Hill, 1961.

Torgovnick, Marianna. *Gone Primitive: Savage Intellects, Modern Lives*. Chicago: University of Chicago Press, 1990.

Tucker, Jennifer. "Voyages of Discovery on Oceans of Air: Scientific Observation and the Image of Science in an Age of 'Balloonacy.'" *Osiris* 11 (1996): 144–176.

U.S. Bureau of the Census. *Historical Statistics of the United States: Colonial Times to 1970*, vol. 1. Washington, DC: U.S. Department of Commerce, 1975.

Viola, Herman J., and Carolyn Margolis, eds. *Magnificent Voyagers: The U.S. Exploring Expedition, 1838–1842*. Washington, DC: Smithsonian Institution Press, 1985.

Whalen, Matthew D., and Mary F. Tobin. "Periodicals and the Popularization of Science in America, 1860–1910." *Journal of American Culture* (1980): 195–203.

Whitley, Richard. "Knowledge Producers and Knowledge Acquirers: Popularization as a Relation between Scientific Fields and Their Publics." In *Expository Science: Forms and Functions of Popularization*, 3–28. Boston: Reidel, 1985.

Wright, John Kirtland. *Geography in the Making: The American Geographical Society, 1851–1951*. New York: American Geographical Society, 1952.

———. "The Open Polar Sea." In *Human Nature in Geography: Fourteen Papers, 1925–1965*, 89–118. Cambridge: Harvard University Press, 1966.

Ziff, Larzer. *Return Passages: Great American Travel Writing, 1780–1910*. New Haven: Yale University Press, 2000.

Index